Kuo Zhang

Characterization and Modeling of the Ratcheting Behavior of the Ferritic-Martensitic Steel P91

Schriftenreihe
des Instituts für Angewandte Materialien
Band 59

Karlsruher Institut für Technologie (KIT)
Institut für Angewandte Materialien (IAM)

Eine Übersicht aller bisher in dieser Schriftenreihe erschienenen Bände finden Sie am Ende des Buches.

Characterization and Modeling of the Ratcheting Behavior of the Ferritic-Martensitic Steel P91

by
Kuo Zhang

KIT Scientific Publishing

Dissertation, Karlsruher Institut für Technologie
KIT-Fakultät für Maschinenbau

Tag der mündlichen Prüfung: 09. Dezember 2015
Referenten: Priv.-Doz. Dr. Jarir Aktaa
Prof. Dr. habil. Jeong-Ha You
Prof. Dr. Oliver Kraft

Impressum

KIT Scientific Publishing

Karlsruher Institut für Technologie (KIT)
KIT Scientific Publishing
Straße am Forum 2
D-76131 Karlsruhe

KIT Scientific Publishing is a registered trademark
of Karlsruhe Institute of Technology.
Reprint using the book cover is not allowed.

www.ksp.kit.edu

Print on Demand 2017 – Gedruckt auf FSC-zertifiziertem Papier

ISSN 2192-9963
ISBN 978-3-7315-0503-7
DOI 10.5445/KSP/1000053265

Acknowledgements

This work was conducted in Institute for Applied Materials – Materials and Biomechanics (IAM-WBM) at Karlsruhe Institute of Technology (KIT).

I am very grateful to my supervisor Prof. Dr.-Ing. Jarir Aktaa, who provided me important guidance during this research. His door is always open whenever I ran into a trouble spot about my research. I would like to thank Prof. Jeong-Ha You and Prof. Oliver Kraft for reviewing my dissertation and providing me valuable suggestions.

Great thanks to Dr. Mario Walter and Mr. Stefan Knaak for being always warmly accessible during the ratcheting experiments. They provided me great assistance for everything in the laboratory. I would like to thank all my friends, Dr. Michael Mahler, Dr. Yiming Wang, Dr. He Pei, Mrs. Dandan Qu for the nice time we spent together and their warmly help.

Great thanks to my wife, Dr. Xiaohui Tang, for her continuous encouragements and great help during these four years. Also great thanks to my little son Shanning Zhang who came to this world in 2014.

Appreciation and thanks should also be given to my parents, Shuiling Zhang and Mei Guo who always fully support my life and research.

Last but not least, I would like to thank all the people who have provided contributions to this research.

Dr. Kuo Zhang
Karlsruhe 29.03.2016

Abstract

The ratcheting behavior of 9% Cr–1% Mo ferritic-martensitic (FM) steel P91 is investigated by uniaxial cyclic loading tests at room temperature and 550 °C. Accumulation rates of strains (ratcheting rates) under multiple loading conditions are recorded to build a database for P91 for further application in Generation IV fission reactors.

Strain-controlled low cycle fatigue (LCF) tests are performed before stress-controlled tests to evaluate the cyclic softening of the material. Afterwards stress-controlled tests are performed both at room temperature and 550 °C, with various peak tensile stresses, mean stresses, stress rates and hold times.

The unconventional asymmetry of stress under strain-controlled LCF tests at room temperature predicted the non-zero ratcheting with zero mean stress, which is equally verified by the subsequent symmetric stress-controlled tests.

A unified visco-plastic deformation model taking into account the complex non-saturating cyclic softening of Reduced Activation Ferritic Martensitic (RAFM) steels is further modified to adapt the ratcheting behavior of P91. It is hereby informed that the current model for RAFM steel fits cyclic softening behavior in strain-controlled LCF tests very well. However, this model strongly overestimates the uniaxial ratcheting rates in stress-controlled tests, because the term for dynamic recovery of kinematic hardening in the current model follows the Armstrong-Frederick dynamic recovery rule.

Based upon further analysis of back stresses, a new constitutive model is proposed. A new dynamic recovery rule is designed to fit the ratcheting rates under multiple loading conditions, including those with smaller stress ratios ($R < -0.8$), larger stress ratios ($R > -0.8$), zero mean stress, various stress rates, and various hold times. Parameter values for various new proposed models are fitted to find the best dynamic recovery rule for kinematic hardening.

Zusammenfassung

Das Ratcheting-Verhalten des 9% Cr–1% Mo ferritisch-martensitischen (FM) Stahls P91 wird durch einachsige zyklische Belastungen bei Raumtemperatur und bei 550°C untersucht. Hierbei werden die Akkumulationsraten der Dehnung (Ratchetingraten) unter verschiedenen Belastungsbedingungen aufgezeichnet, um die Datenbank für P91 in Bezug auf die Anwendung in der Generation IV Spaltungsreaktoren zu erweitern.

Zu Beginn werden dehnungsgesteuerte LCF-Versuche und anschließend spannungsgesteuerte Versuche durchgeführt, um die zyklische Entfestigung des Materials zu bewerten. Die spannungsgesteuerten Versuche werden sowohl bei Raumtemperatur als auch bei 550°C unter Variation der Spannungsspitzen, der Mittelspannungen, der Spannungsraten und der Haltezeiten durchgeführt. Die untypische Asymmetrie der Spannung bei den dehnungsgesteuerten LCF-Versuchen bei Raumtemperatur ist die Ursache für das Auftreten von Ratcheting ohne Mittelspannung. Außerdem wird dies durch die folgenden symmetrischen spannungsgesteuerten Versuchen bestätigt.

Um das Ratcheting-Verhalten von P91 zu beschreiben, wird ein viskoplastisches Verformungsmodell unter Berücksichtigung der komplexen nicht sättigenden zyklischen Entfestigung der Reduced Activation Ferritic Martensitic (RAFM) Stähle weiter modifiziert. Es wird gezeigt, dass das aktuelle Modell für RAFM Stähle in der Lage ist, das zyklische Entfestigungsverhalten bei spannungsgesteuerten LCF-Versuchen sehr gut zu beschreiben. Allerdings überschätzt dieses Modell offensichtlich die einachsigen Ratchetingraten in spannungsgesteuerten Versuchen. Verantwortlich ist die dynamische Erholung der kinematischen Verfestigung im aktuellen Modell, die auf das Modell von Armstrong-Frederick zurückzuführen ist.

Es wurde ein neues konstitutives Modell, welches auf Untersuchungen der Rückspannung basiert, entwickelt. Zur besseren Beschreibung des Ratchetings wurde ein neuer dynamischer Erholungsterm verwendet, der die unterschiedlichen Belastungsbedingungen berücksichtigt. Typische Belastungsbedingungen sind beispielsweise kleine ($R<-0{,}8$), und große ($R>-0{,}8$) Spannungsverhältnisse, keine Mittelspannung, Variation der Spannungsraten und unterschiedliche Haltezeiten. Es wurden Parametersätze für die verschiedenen vorgeschlagenen neuen Modelle bestimmt, mit denen die dynamische Erholung der kinematischen Verfestigung gut beschrieben werden kann.

Contents

Figure List

Table List

Abbreviation

AF	Armstrong-Frederick
AS	Aktaa Schmitt
BS	Back Stress
CEA	French Alternative Energies and Atomic Energy Commission
CERN	European Organization for Nuclear Research
FM	Ferritic-Martensitic
LCF	Low Cycle Fatigue
MATTER	MATerial TEsting and Rules
OW	Ohno-Wang
OAK	Ohno-Abdel-Karim
PAGB	Prior Austenite Grain Boundaries
RT	Room Temperature
RAFM	Reduced Activation Ferritic Martensitic
YT	Yaguchi-Takahashi

Notation

ε	Strain
$\Delta\varepsilon$	Strain range
ε_{peak}	Peak tensile strain /maximum strain
ε_{min}	Minimum strain
ε^{in}	Inelastic strain
$\Delta\varepsilon^{in}$	Inelastic strain range
$\varepsilon_r / \varepsilon_{mean}$[1]	*Ratcheting strain/mean strain*
ε'_r[2]	Ratcheting rate
σ	Stress
$\sigma_{peak} / \sigma_{max}$[3]	Peak stress/peak tensile stress/maximum stress
σ_{min}	Minimum stress
σ_{mean}	Mean stress
σ_a	Stress amplitude
σ_{eng}	Engineering stress
σ_{true}	True stress
$\dot{\sigma}$	Stress rate
R	Stress ratio
Ω	Back stress/ Kinematic hardening

1 Ratcheting strain is mean value of maximum and minimum strain in one cycle, hence the term "mean strain" and "ratcheting strain" is identical in the topic of ratcheting.

2 Ratcheting rate is the increase/decrease of mean strain of one cycle to the one in the previous cycle, namely $\acute{\varepsilon}_r = \frac{\Delta\varepsilon_r}{\Delta N}$.

3 In case of confusion, the default meaning of peak stress is peak tensile stress. Since in all experiments mentioned in literature and performed in the current PhD program, the maximum stresses are tensile stresses, therefore, the term "maximum stress", "peak stress" and "peak tensile stress" are synonyms in this paper. On the other hand, in spite of a few cases where minimum stresses are zero or tensile stresses, minimum stresses are generally compressive stresses. Therefore, the term "peak compressive stress" is the same as "minimum stress" if without extra explanation.

1 Introduction

The task of the current PhD program is a part of the MATerial Testing and Rules (MATTER) project, which is a material research program for the construction of a European Generation IV reactor. Currently available tests and evaluation standards are not sufficient to predict the structural material behavior under the operational conditions of the LFR ETPP (Lead Fast Reactor European Technology Pilot Plant) MYRRHA and SFT Prototype ASTRID, which are two prototype European Gen. IV reactors.

The scope of the MATTER project is to contribute to covering the existing gaps by pointing out methodologies, recovering existing experiences and performing experiments. One of the goals of the MATTER project is to provide the design rules for 9Cr–1Mo ferritic-martensitic (FM) steel.

For the application of 9Cr–1Mo FM steel in the construction of a nuclear power plant, not only a wide database of its mechanical characteristics is required, but also new rules need to be set, with which reliable construction planning can be done according to the characteristics of the steel. Although FM steels are preferable to austenitic steels in reactor construction owing to their lower swelling under radiation, they still have a negative side that they show cyclic softening. Cyclic softening plays a major role in ratcheting, in the sense that the strain ranges of hysteresis loops can increase cycle by cycle owing to softening and it accelerates the ratcheting. The current criteria regarding the influence of cyclic loading on ratcheting are limited since they are mainly developed for materials showing cyclic hardening. These criteria, if used without further improvement, cannot be applied to the constructions with 9Cr–1Mo FM steels.

Although many investigations have been carried out on 9Cr–1Mo FM steels, the ratcheting behavior has not been extensively studied. This is because, firstly, a variety of unique material characteristics were not reported in every study on this type of steel, such as cyclic softening and asymmetry of material strength under tension and compression; secondly, a thorough study on the influence of cyclic softening on the ratcheting is lacking; thirdly, the material responses under a variety of loading conditions have not been extensively evaluated. On the other hand, although various models have been proposed to simulate the ratcheting behavior of the material, a simple, robust and convincing model is still missing, owing to excessive parameters in the previously developed models and the lack of tests under various loading conditions to verify the simulation ability of these models.

In this PhD program, uniaxial material behavior of 9Cr–1Mo steel P91 at room temperature (RT) and at 550 °C is studied. The experimental data from both strain-controlled low cycle fatigue (LCF) tests and stress-controlled ratcheting tests are collected to build a database for P91. Cyclic softening is evaluated according to the data collected from strain-controlled tests and the factors influencing ratcheting, including peak stress (σ_{peak}), mean stress (σ_{mean}), stress rate ($\dot{\sigma}$) and hold time are evaluated according to the data collected from the stress-controlled tests.

The constitutive model proposed by Aktaa and Schmitt [1] describes typical cyclic softening for Reduced Activation Ferritic Martensitic (RAFM) steels. Hence, this model was also named the RAFM model, which has been proved to have good simulation ability for strain-controlled LCF tests on EUROFER 97 and F82H mod.

Based on the developed database built in this work, the simulation ability of RAFM model for P91 is tested. Especially the ability to simulate the ratcheting strain (ε_r) under a variety of stress-controlled loading conditions is tested. A new constitutive model is proposed based on the RAFM model to simulate the uniaxial isothermal behavior of P91 at both RT and 550 °C.

In Chapter 1, a variety of previous studies are reviewed, including those focusing on ratcheting of various materials, modeling approaches for ratcheting, and characteristics such as cyclic softening of FM steel.

In Chapter 1, the specimens and experimental facility are presented. The experiment planning at RT and 550 °C is presented, including both strain- and stress-controlled tests.

In Chapter 1 experiments at RT are illustrated with diagrams and initial analysis. The effects of various influencing factors on ratcheting are presented in separate sections.

Following the same structure, Chapter 1 presents the experiments at 550 °C.

Chapter 1 presents the detailed process of the development of the new constitutive model, starting from the modeling criteria. The simulation ability of the RAFM model and some other proposed models are tested for P91. A final designed model is chosen and its simulation ability is verified by comparing the model and material responses under multiple loading conditions.

In Chapter 7, the experiment and simulation results are further discussed.

Chapter 7 summarizes the current work and proposes several suggestions for future research on ratcheting of FM steel.

2 State of the Art

2.1 Ratcheting Effect

The word "ratchet" is the name of a mechanical device that allows continuous linear or rotary motion in only one direction. In material science, the term "metaphorical ratcheting effect" is used synonymously with "progressive deformation" [2], which means that the mean strain (ε_{mean}) (arithmetic mean of maximum and minimum strain during one loading cycle) accumulates only in one direction when the structure is subjected to asymmetric cyclic loading. Ratcheting is known as "cyclic creep", owing to the similar feature of "monotonic increase of strain" as "creep". However, creep deformations generally only become obvious at a temperature above approximately 30% of the melting point, while "cyclic creep" or ratcheting is already observable at much lower temperatures. Further, creep is a result of long-term stress. Therefore, creep is a "time-dependent" deformation, while ratcheting can be either time-dependent or time-independent. Note that the term "mean strain (ε_{mean})" and "ratcheting strain (ε_r)" is identical in the topic of ratcheting.

The ratcheting behavior can be distinguished into material ratcheting and structural ratcheting. Material ratcheting occurs without structural effects, assuming the stress is distributed homogeneously in a structure. It is a purely material-related effect, which can be analytically modeled with constitutive equations. Structural ratcheting, on the other hand, can occur even if there is no material ratcheting. It happens due to inhomogeneity of the state of stress in a structure [2]. In the current work, only material ratcheting is taken into account, so in this report, the word "ratcheting" means only material ratcheting.

Ratcheting test is performed under stress-controlled cyclic loading during which the hysteresis loops do not close. As a result of non-closed hysteresis loops, the ε_{mean} during each loading cycle is different from that in the previous cycle. In most cases, ε_{mean} accumulates in the direction of σ_{mean} (arithmetic mean of maximum and minimum stress during one loading cycle).

Note that there is a distinction between the term "strain accumulation" $\varepsilon^{acc}{}_{ij} = \int \dot{\varepsilon}^{pl}{}_{ij} dt$ and "accumulated plastic strain" $p = \int \left(\frac{2}{3}\dot{\varepsilon}^{pl}{}_{ij}\dot{\varepsilon}^{pl}{}_{ij}\right)^{1/2} dt$ [2]. Strain accumulation $\varepsilon^{acc}{}_{ij}$ is zero in a closed hysteresis loop while accumulated plastic strain p can only increase monotonically. Accumulated plastic strain p is used in various constitutive theories [3-7].

One of the earliest observations of ratcheting was reported for 1100 aluminum, which showed shifting of hysteresis loops in the presence of mean stress [8]. Researchers have found ratcheting effect on a wide range of materials, including:

- Austenitic steels such as 316L [9-11], 316LN [12], 304 [13-20], 304L [21], 304LN [22] (304L means "low carbon", 304LN means "low carbon and high nitrogen" [23]);
- Ferritic steels [24-28];
- Carbon steels (CS) 40Cr [29], 42CrMo [18, 30, 31], 16MnR[32], X42 and X56 [33], 20 CS [34, 35], 45 CS [36];
- Other steels such as SA333C–Mn steels [37], Interstitial-Free Steel [38];
- Other metallic materials, such as zirconium alloys [39], pure titanium [40], titanium alloys [41], NiTi shape memory alloy [42, 43];
- Ceramic matrix composites [44]; and
- Polymers such as polytetrafluoroethylenes (PTFE) [45], polypropylene [46], polyacetal/polyoxymethylene[47], epoxy resin[48] etc.

Some of the research on ratcheting of a variety of steels is reviewed in the following.
Although ratcheting tests are supposed to be performed under stress-controlled cyclic loading, strain-controlled ratcheting tests were also performed [9, 10], which are known as “cyclic tension tests”. The imposed strain was a combination of an alternate strain and a mean strain, as shown in Fig. 2.1).

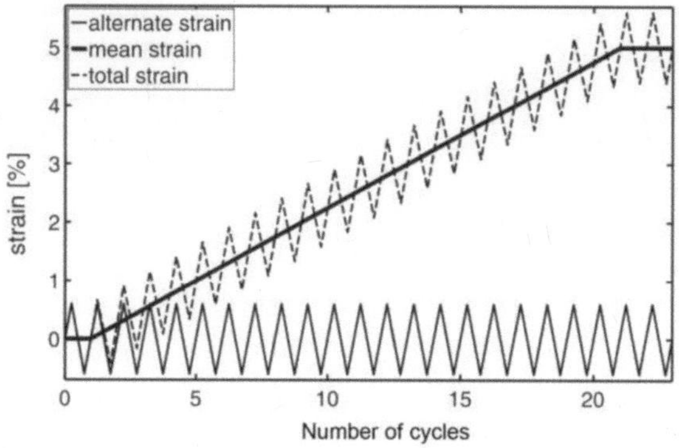

Fig. 2.1: Generic strain path for a cyclic tension test where the total strain path is a superposition of mean and alternate strain [10].

Such strain-controlled ratcheting tests were performed on AISI 316L austenitic stainless steel at RT and 200 °C [10]. It was found that the material response was characterized by an additional hardening and a non-zero σ_{mean}, and the authors suggested that the response was a superposition of two mechanisms: a cyclic one (hardening corresponding to LCF behavior) and a monotonic one (hardening owing to drifting of the ε_{mean}). So-called “fatigue/creep-fatigue tests with superimposed strain” were carried out on mod. 9Cr–1Mo steel at 550 and 450 °C [28], which were essentially the same as the “strain-controlled ratcheting tests”. However, stress-controlling still plays a large part in ratcheting research. The other ratcheting research discussed in this chapter is all stress-controlled if without extra explanation.

Ratcheting tests can be distinguished into uniaxial and multiaxial tests. Although uniaxial tests require a relatively simple experimental facility, multiaxial tests are closer to real working conditions in the sense that real components are generally under multiaxial loading. Specimens for multiaxial ratcheting tests are tubular with typical outer/inner diameters of 13/10mm [26], 15/12mm [15, 17, 19], 16/13mm [11, 31, 35], or 22/18mm[49]. To verify the multiaxial constitutive models, not only axial stresses but also shear stresses were placed on tubular specimens. Various steels have been tested under multiaxial loading, such as mod. 9Cr–1Mo steel [26], 304 [15, 17, 19], 304L[50], 316L [11] , carbon steel 42CrMo [31], carbon steel 20 [35], U71Mn rail steel [51] and 1Cr18Ni9Ti stainless steel [49] . A variety of loading paths used in multiaxial tests are illustrated in Fig. 2.2.

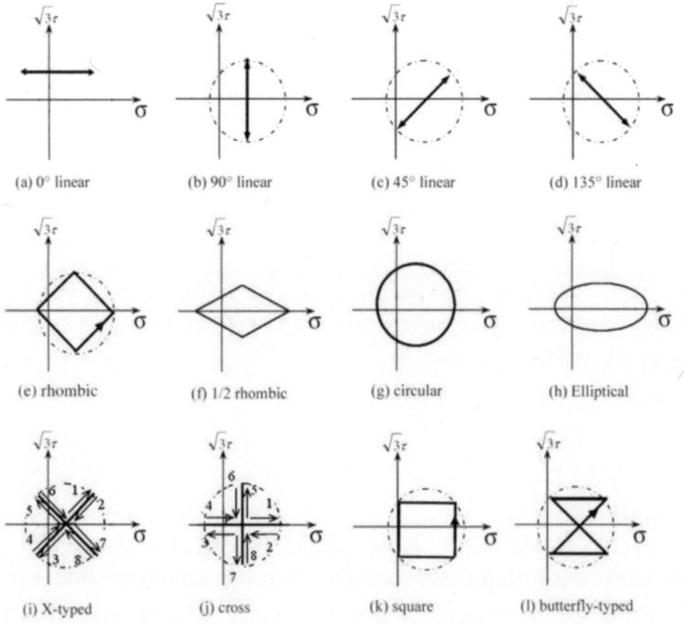

Fig. 2.2: Shapes of some loading paths used in the multiaxial stress-controlled cycling tests.[11, 15, 17, 31, 35, 49, 50]

However, for reasons of simplicity and restriction of testing facility, uniaxial ratcheting tests still play a large role in ratcheting research.

Specimens in uniaxial ratcheting tests are mostly in the shape of a solid cylinder; however, cylindrical shell of stainless steel 304L has been also subjected to uniaxial loading [21], since the shell form has a wide range of applications in industry owing to its lightweight and high strength. The ratcheting behavior of cylindrical shell was noted at RT. Although the shells mentioned in [21] and the tubular specimens for multiaxial ratcheting tests [11, 15, 17, 19, 26, 31, 35, 49] had similar shape, the shells in [21] had an outer diameter of 42mm and shell thickness 1.5mm, with the diameter-to-thickness ratio of 28, while in e.g. [11], this ratio of the tubular specimen was only 10. With larger diameter-to-thickness ratio, the structure was more susceptible to buckling. Note that the uniaxial ratcheting was studied together with buckling behavior of the shells [21].

The failure modes under uniaxial cyclic loading were investigated with carbon steel 45 at RT [36]. Strain-controlled tests were firstly performed to fit the parameters in the Coffin–Manson formula [52-54] with cycle number to the failure of the material (N_f). Then ratcheting behavior was studied in stress-controlled tests. It was found that, in strain-controlled tests, and in stress-controlled tests performed with high σ_a and small σ_{mean}, the failure mode was fracture, while in stress-controlled tests performed with relatively small σ_a and large σ_{mean}, the failure mode was ductile localized necking. Therefore, the fatigue damage D_f was relatively larger with larger σ_a and smaller σ_{mean}, while D_f was relatively smaller with relatively smaller σ_a and larger σ_{mean} [36].

The uniaxial and non-proportionally multiaxial ratcheting behavior of austenitic steel 304 was researched under asymmetrical stress-controlled cyclic loading with variable σ_a and σ_{mean}, loading paths, and loading histories. A phenomenon named "dynamic strain aging" was found in the temperature range of 400–600 °C, in which much greater cyclic hardening and less ratcheting were observed than at RT [15]. The explanation for this phenomenon was that the interactions of dislocation and point defect were significantly active, which resulted in a remarkable enhancement of deformation resistance, hence the cyclic hardening was greater. This phenomenon was not reported for the other steels.

In most ratcheting research [3-5, 9, 14, 16, 24, 26, 30, 36], minimum stress in each hysteresis loop was compressive. On the contrary, uniaxial ratcheting tests with tensile minimum stresses were carried out on ferritic steel X12CrMoWVNbN10-1-1 with various hold times and stress ratios at a temperature of 600 °C [25, 27]. Because the minimum stress in each cycle was tensile, the hysteresis loops were always within the tensile range. The total accumulated strain was decomposed into accumulated ratcheting strain (partial inelastic strain formed during stress-changing process) and accumulated creep strain (strain increase during hold time). The so-called shakedown behavior of ratcheting was observed on specimens subjected to a relatively long hold time (i.e. 5 and 20min), which meant the partial inelastic strain formed during the stress-changing process of each cycle continuously decreased until no ratcheting was observed [25, 27].

On the other hand, when the hold time was less than 5min, the total accumulated strain was mainly composed of the increased ratcheting strain owing to the inelastic creep recovery. For longer hold times (10, 20 or 30min) however, the accumulated creep strains were the controlling mechanism of deformation [25, 27].

Generally speaking, ratcheting happens with a non-zero σ_{mean}, or in other words, asymmetric stress-controlled loading, while positive ratcheting (strain accumulation in the direction of tensile stress) with zero σ_{mean} on mod. 9Cr–1Mo steel at 550 °C was also observed [7, 26]. This behavior was named “unconventional ratcheting”. It was suggested that the reason for this unconventional behavior had something to do with hydrostatic pressure and this suggestion was used in the construction of the constitutive model [7, 26]. The corresponding modeling approach is discussed in detail in Section 2.2.

In another study on ratcheting of mod. 9Cr–1Mo steel, a phenomenon named "progressive deformation instability" was observed at 600 °C [24]. Fig. 2.3 and Fig. 2.4 show diagrams of ε_r versus number of cycles. Fig. 2.3 shows the results with the same σ_{mean} and different σ_a, while Fig. 2.4 shows the results with the same σ_a and different σ_{mean}. It was clear that ratcheting rate $\acute{\varepsilon}_r$ (change of mean strain/ratcheting strain per cycle) increased with increasing σ_a (with the same σ_{mean}) and increased with increasing σ_{mean} (with the same σ_a). The sudden changes of $\acute{\varepsilon}_r$ were the so-called "progressive deformation instability", which was explained as being due to severe cyclic softening characteristic of the tested material [24]. A detailed discussion of cyclic softening is provided in Section 2.5.

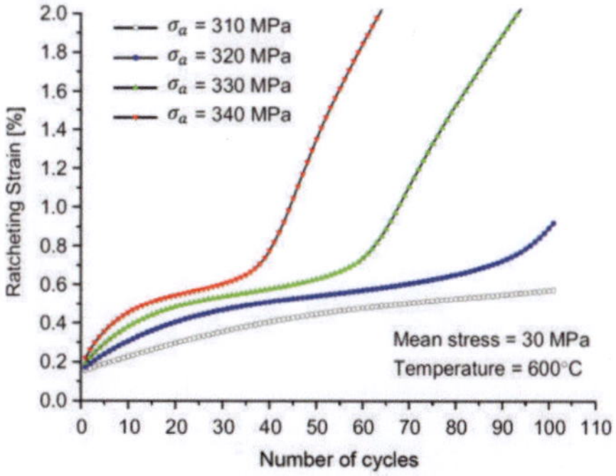

Fig. 2.3: Effects of the applied stress amplitude on progressive deformation instability owing to the cyclic softening of the mod. 9Cr–1Mo steel [24].

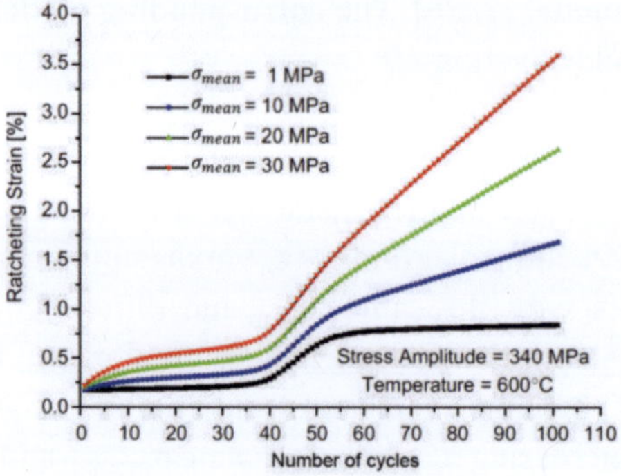

Fig. 2.4: Effects of the mean stress on progressive deformation instability owing to the cyclic softening of the mod. 9Cr–1Mo steel [24].

However, such "progressive deformation instability" was not mentioned for mod. 9Cr–1Mo steel at 550 °C [7, 26] while only an increase of strain range was observed in tests at 550 °C, as shown in Fig. 2.5.

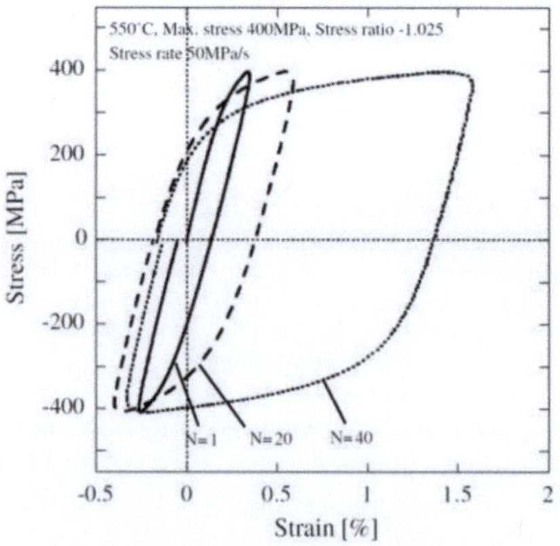

Fig. 2.5: Stress–strain hysteresis loops under stress ratio of -1.025 [26].

Compared to austenitic steels, ratcheting research on ferritic steels, especially on mod. 9Cr–1Mo steel is relatively sparse. Currently, no reports about ratcheting behavior of this material at RT can be found in the literature, and some research at 550 °C is still doubtful. For instance, the ultimate tensile strength measured in SCK•CEN [55] at 550 °C was 374 MPa, however according to [26], a group of ratcheting tests were performed with ε_{peak}= 400 MPa at 550 °C which would be impossible if they were testing on the same material, since both materials mentioned in [26] and [55] were mod. 9Cr–1Mo FM steels. On the other hand, the positive ratcheting in the vicinity of zero σ_{mean} reported in [26] also requires further verification. Further, the progressive deformation instability induced by cyclic softening at 600 °C [24] should be checked at 550 °C.

2.2 Modeling of Ratcheting

One of the main aims in research on ratcheting is to give a better model of the visco-plasticity characteristic of materials with better prediction of the ratcheting effect. Most work in recent decades [3, 4, 6, 7, 14, 30, 56, 57] is based on the Chaboche model [5, 58-61], the Armstrong–Frederick (AF) rule of dynamic recovery of kinematic hardening [62], and the Ohno–Wang (OW) model [3, 4].

The Chaboche model was a so-called unified deformation model that described visco-plasticity without the separation in time-dependent creep and time-independent plasticity [1, 63]. "Standard" constitutive models, which were usually used in finite-element codes, decomposed the total strain into elastic, plastic, creep and anelastic contributions, while the "unified" model considered creep and plasticity as arising from the same dislocation source [62].

The basic visco-plastic equations in the Chaboche model are as follows:

$$\dot{\boldsymbol{\varepsilon}}^{in} = \frac{3}{2}\langle\frac{\Sigma_{eq}-K}{Z}\rangle^{n}\frac{\boldsymbol{\Sigma}}{\Sigma_{eq}} \quad (2\text{-}1)$$

$$\boldsymbol{\Sigma} = \boldsymbol{s} - \boldsymbol{\Omega}, \tag{2-2}$$

$$\Sigma_{eq} = \sqrt{\frac{3}{2}\boldsymbol{\Sigma}:\boldsymbol{\Sigma}} \tag{2-3}$$

$$\boldsymbol{s} = \boldsymbol{\sigma} - \frac{1}{3}trace(\boldsymbol{\sigma})\mathbf{1} \tag{2-4}$$

$$\boldsymbol{\Omega} = \sum_{i=1}^{M} \boldsymbol{\Omega}_i \tag{2-5}$$

$\langle\,\rangle$ indicated the Macaulay bracket, which operated as $\langle x \rangle = 0$ when $x < 0$ and $\langle x \rangle = x$ when $x \geq 0$. $\boldsymbol{\Omega}$ represented the back stress (BS). M meant that the BS was composed of M sub-components. K, Z, and n were material and temperature-dependent parameters.
To avoid confusion, all notation in this work follows the form as in Aktaa and Schmitt [1]. A comparison between notations in Aktaa and Schmitt [1] and notations in other reports can be found in Table 2.1.

[1]	[5, 58]	[3, 4, 6]
$\boldsymbol{s}$	$\boldsymbol{\sigma}'$	$\boldsymbol{s}$
$\boldsymbol{\Omega}$	$\boldsymbol{X}$	$\boldsymbol{a}$
Σ_{eq}	$J(\boldsymbol{\sigma} - \boldsymbol{X})$	σ_{eff}
$\boldsymbol{\varepsilon}^{in}$	$\boldsymbol{\varepsilon_p}$	$\boldsymbol{\varepsilon}^{p}$
H_i	C_i	$h_i = \zeta_i \cdot r_i$
C_i	γ_i	ζ_i

Table 2.1: Comparison of notations in different reports.

From the microstructural point of view, the term "back stress" is defined as follows: The acting stress on the leading dislocation in a pile-up is the acting stress on the glide plane multiplied by the number of dislocations in the pile-up. A similar stress, but with opposite sign, opposes the operation of the generator, in the form of a "back stress"[64]. In the theory of solid mechanics concerning flow laws, BS is referred to as "kinematic hardening variable". It is also known as the "microstress component" [62].

Armstrong and Frederick [62] provided a type of equation of BS which can be simplified as follows:

$$\dot{\boldsymbol{\Omega}}_i = \frac{2}{3} H_i \dot{\boldsymbol{\varepsilon}}^{in} - C_i \boldsymbol{\Omega}_i \dot{p} - R_i |\boldsymbol{\Omega}_i|^{m_i - 1} \boldsymbol{\Omega}_i \tag{2-6}$$

with

$$\dot{p} = \sqrt{\frac{2}{3} \dot{\boldsymbol{\varepsilon}}^{in} : \dot{\boldsymbol{\varepsilon}}^{in}} \tag{2-7}$$

The second and the last term in eq. (2-6) represent the dynamic recovery and static recovery of the back stress component $\boldsymbol{\Omega}_i$, respectively. H_i, C_i, R_i and m_i are the material and temperature dependent parameters. This is commonly referred to as the Armstrong–Frederick rule of kinematic hardening (AF rule).

Note that the static recovery term was already introduced in the original Armstrong–Frederick report [62], but with a linear dependency of BS. This term was not included in many modeling approaches [3, 4, 6, 51, 58].

A well-known disadvantage of the AF rule is that, it predicts too much accumulated strain under non-symmetric loading conditions [6, 62]; in other words, constitutive models with the AF rule predict too much ratcheting.

Improved rules were developed to avoid the defect of the AF rule on ratcheting prediction by, for example, introducing a power function of the BS in the dynamic recovery term as follows:

$$\dot{\boldsymbol{\Omega}}_i = \frac{2}{3} H_i \dot{\boldsymbol{\varepsilon}}^{in} - \frac{{C_i}^2}{H_i} [J(\boldsymbol{\Omega}_i)]^{m-1} \boldsymbol{\Omega}_i \dot{p} \tag{2-8}$$

where $J(\boldsymbol{x})$ represents the von Mises invariant $\left(\frac{3}{2}\boldsymbol{x}':\boldsymbol{x}'\right)^{1/2}$. $\boldsymbol{x}'$ is the deviator of $\boldsymbol{x}$, as in eq.(2-4). Note that there was no static recovery term in eq.(2-8) [58].

The Ohno–Wang model includes Ohno–Wang model I (OW I) and model II (OW II). It was assumed that the dynamic recovery of BS is activated fully only when its magnitude $\overline{\boldsymbol{\Omega}}_i$ attains a critical value, resulting from the energy required for cross slip [3, 4]. In OW I, the critical state of dynamic recovery is represented by a surface $f_i = 0$:

$$f_i = {\overline{\boldsymbol{\Omega}}_i}^2 - {r_i}^2 \tag{2-9}$$

with parameter $r_i = \frac{H_i}{C_i}$ and

$$\overline{\boldsymbol{\Omega}}_i = \sqrt{\frac{3}{2}\boldsymbol{\Omega}_i : \boldsymbol{\Omega}_i} \tag{2-10}$$

The equation for BS in OW I is as follows:

$$\dot{\boldsymbol{\Omega}}_i = \frac{2}{3} H_i \dot{\boldsymbol{\varepsilon}}^{in} - Hs(f_i) C_i \boldsymbol{\Omega}_i \dot{\lambda}_i \tag{2-11}$$

with

$$\dot{\lambda}_i = \langle \dot{\boldsymbol{\varepsilon}}^{in} : \boldsymbol{k}_i \rangle \tag{2-12}$$

where $\boldsymbol{k}_i$ denotes the direction of $\boldsymbol{\Omega}_i$.

$$\boldsymbol{k}_i = \frac{\boldsymbol{\Omega}_i}{\overline{\Omega}_i} \tag{2-13}$$

Fig. 2.6 illustrates the evolution of BS component $\boldsymbol{\Omega}_i$. $\dot{\lambda}_i$ takes the form of eq. (2-12) to keep $\boldsymbol{\Omega}_i$ in the critical state $f_i = 0$.

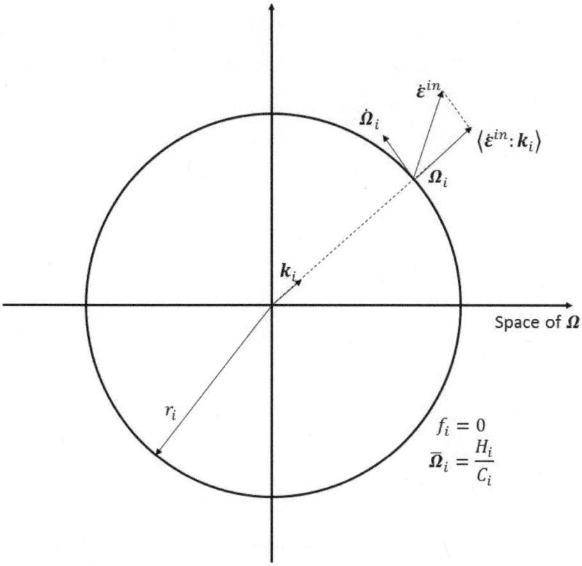

Fig. 2.6: Evolution of $\boldsymbol{\Omega}_i$ on the surface of $f_i = 0$ [3]

Hs denotes the Heaviside step function, which operates as $Hs(x) = 0$ when $x < 0$ and $Hs(x) = 1$ when $x \geq 0$.

However, the simulated hysteresis loops with OW I are piecewise linear, without everywhere-differentiability and expressed no uniaxial ε_r [3, 4].

In OW II, the Heaviside step function in OW I is replaced with an everywhere-differentiable term $\left(\frac{\bar{\boldsymbol{\Omega}}_i}{r_i}\right)^{m_i}$. Equation for BS in OW II is as follows:

$$\dot{\boldsymbol{\Omega}}_i = \frac{2}{3} H_i \dot{\boldsymbol{\varepsilon}}^{in} - \left(\frac{\bar{\boldsymbol{\Omega}}_i}{r_i}\right)^{m_i} C_i \boldsymbol{\Omega}_i \dot{p} \tag{2-14}$$

in which m_i is the material- and temperature-dependent parameter. When $m_i \to \infty$, eq. (2-14) reduced into eq. (2-11).

Fig. 2.6 in [3] shows the comparison between the OW I, II and AF model. In both OW I and II, the magnitudes of the dynamic recovery terms were minimized, while OW II avoided the non-differentiable corner when $\bar{\boldsymbol{\Omega}}_i \to {}^{H_i}/_{C_i}$.

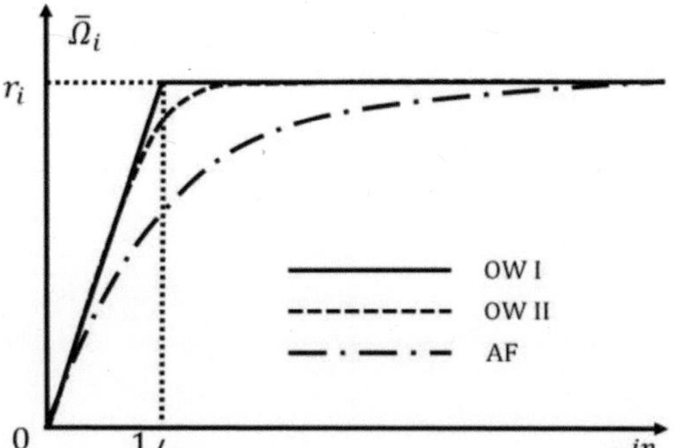

Fig. 2.7: Change of $\bar{\Omega}_i$ under uniaxial tensile loading [3]

Both OW I and II should have four or eight BS components to fit the experimental loop shapes and ε_r, which required too much effort in fitting the parameters. On the other hand, the two examples in [4] to verify the simulation ability of OW model were with σ_{peak} = 400 MPa & σ_{mean} =100 MPa and with σ_{peak} = 400 MPa & σ_{mean} =150 MPa, respectively, in which ε_r in the material responses were too small (no more than 0.5%). Hence, these two verifications were not persuasive. Another modeling approach was to combine the AF rule into the OW model in the evolution equation for dynamic recovery, which can be referred to as Ohno–Abdel–Karim (OAK) model [6]:

$$\dot{\boldsymbol{\Omega}}_i = \frac{2}{3}H_i\dot{\boldsymbol{\varepsilon}}^{in} - \mu_i C_i \boldsymbol{\Omega}_i \dot{p} - Hs(f_i)C_i\langle\dot{\lambda}_i\rangle\boldsymbol{\Omega}_i \tag{2-15}$$

with

$$\dot{\lambda}_i = \dot{\boldsymbol{\varepsilon}}^{in}:\frac{\boldsymbol{\Omega}_i}{r_i} - \mu_i\dot{p} \tag{2-16}$$

The parameter μ_i combines the AF rule into the OW model in eq. (2-15): When μ_i=0, eq. (2-15) was the same in OW I as eq. (2-11). If μ_i=1, it was the same in the AF rule as eq. (2-6), in spite of the term for static recovery.

To determine the parameters H_i and C_i, the relation of BS $\boldsymbol{\Omega}$ and inelastic strain $\boldsymbol{\varepsilon}^{in}$ was linearized and eq. (2-15) was correspondingly reduced to OW I (μ_i = 0), which was perfectly linear. As illustrated in Fig. 2.8, multiaxial case was simplified into the uniaxial case and the curve of Ω-ε^{in} was linearized into 3 linear sections with corners (M = 3) [6].

As $\mu_i \rightarrow 0$,

$$C_i = \frac{1}{\varepsilon_{(i)}^{in}} \tag{2-17}$$

$$H_i = C_i\left[\frac{\Omega_{(i)} - \Omega_{(i-1)}}{\varepsilon_{(i)}^{in} - \varepsilon_{(i-1)}^{in}} - \frac{\Omega_{(i+1)} - \Omega_{(i)}}{\varepsilon_{(i+1)}^{in} - \varepsilon_{(i)}^{in}}\right]\varepsilon_{(i)}^{in} \tag{2-18}$$

where $\varepsilon_{(0)}^{in}$and $\Omega_{(0)}$=0.

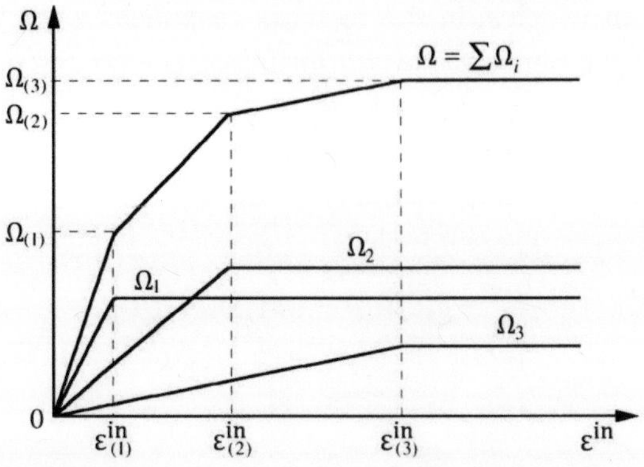

Fig. 2.8: Change of BS and its sub-components under uniaxial tensile loading in the case of $\mu_i = 0$, $(M = 3)$ [6].

The BS was decomposed into eight components in [6] to simulate the material responses of mod. 9Cr–1Mo steel, which meant eight groups ($M = 8$) of H_i and C_i should be determined. This OAK model developed in [6] was employed in [17, 20] to simulate uniaxial and multiaxial ratcheting of stainless steel 304.

The equation of BS was further modified into eq. (2-19) in another modeling approach [51].

$$\dot{\boldsymbol{\Omega}}_i = \frac{2}{3} H_i \dot{\boldsymbol{\varepsilon}}^{in} - \mu_i C_i \boldsymbol{\Omega}_i \dot{p} - Hs(f_i) C_i (1 - \mu_i) \dot{p} \boldsymbol{\Omega}_i \tag{2-19}$$

The term $\langle \dot{\boldsymbol{\varepsilon}}^{in} : \frac{\boldsymbol{\Omega}_i}{r_i} - \mu_i \dot{p} \rangle$ in eq. (2-15) was changed into $(1 - \mu_i)\dot{p}$ in eq. (2-19). This model proposed in [51] can be referred to as the Kang model, which was employed in [30] to simulate uniaxial ratcheting of 42CrMo steel and in [65] to study uniaxial ratcheting and to predict multiaxial ratcheting of SiC_p/6061Al composites.

Note that the static recovery term in AF rule (see eq. (2-6)) was not included in the OW I and II models, the OAK model or the Kang model.

In another modeling approach, the cyclic softening was taken into account in the simulation of the ratcheting behavior of mod. 9Cr–1Mo steel at 550 °C [7]. The simulation of cyclic softening in [7] is discussed in detail in Section 2.5.

The model proposed in [7] can be referred to as the Yaguchi–Takahashi (YT) model. OW I model was applied in the YT model with the addition of the static recovery term of BS. The equation for BS in the YT model is as follows:

$$\dot{\boldsymbol{\Omega}}_i = \frac{2}{3} H_i \dot{\boldsymbol{\varepsilon}}^{in} - H s(f_i) C_i \left\langle \dot{\boldsymbol{\varepsilon}}^{in} : \frac{\boldsymbol{\Omega}_i}{\overline{\boldsymbol{\Omega}}_i} \right\rangle \boldsymbol{\Omega}_i - R_i |\boldsymbol{\Omega}_i|^{m_i - 1} \boldsymbol{\Omega}_i \tag{2-20}$$

$$f_i = \overline{\boldsymbol{\Omega}}_i^{\,2} - r_i^{\,2} \tag{2-9}$$

Although the original OW I expressed no ε_r, the term of static recovery in eq. (2-20) yielded a non-zero ε_r, which was confusing because the ratcheting was supposed to be only owing to the static recovery of BS in YT model. Moreover, this model overestimated the ratcheting particularly with larger σ_{mean}, as shown in Fig. 2.9.

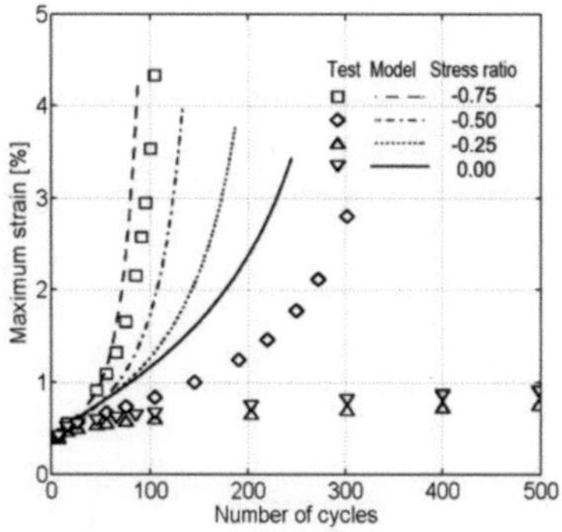

Fig. 2.9: Simulations of uniaxial ratcheting under stress ratio conditions between -0.75 and 0 by proposed constitutive model, with $\sigma_{\max} = 400$MPa, stress rate ± 50MPa/s[7].

As mentioned in Section 2.1, positive ratcheting with zero σ_{mean} was reported in [26]. Hydrostatic pressure was suggested to be the reason for this unconventional ratcheting behavior. Accordingly, the Chaboche model was modified with the addition of $J_m(\boldsymbol{\sigma})$ [7].

$$J_m(\boldsymbol{\sigma}) = \eta|tr(\boldsymbol{\sigma})|^l sgn(trace(\boldsymbol{\sigma})) \tag{2-21}$$

with $sgn(x)$ works as

$$sgn(x) = \begin{cases} 1 & if\ x > 0 \\ 0 & if\ x = 0 \\ -1 & if\ x < 0 \end{cases}$$

Eq. (2-1) in the Chaboche model was modified to the following equation:

$$\dot{\boldsymbol{\varepsilon}}^{in} = \frac{3}{2}\langle\frac{\Sigma_{eq}-K+J_m(\sigma)}{Z}\rangle^n\frac{\boldsymbol{\Sigma}}{\Sigma_{eq}} \tag{2-22}$$

Consequently, the value of $\dot{\boldsymbol{\varepsilon}}^{in}$ was larger under tensile stress than compressive stress, since $J_m(\boldsymbol{\sigma})$>0 when $tr(\boldsymbol{\sigma})$>0. As a result, under symmetric cyclic loading, the rate of inelastic strain $\boldsymbol{\varepsilon}^{in}$ was larger in the tensile part of each hysteresis loop than in the compressive part, hence the loops were not closed and the ε_{mean} during each loading cycle was larger than that of the earlier cycle, even when the peak tensile stresses were equal to the peak compressive stresses. In other word, the model yielded positive ratcheting with zero σ_{mean}.

Instead of four, eight or even 12 BS components as in [4, 6, 7, 51], a further modeling approach reduced the number, leaving merely three BS components, which reduced the complexity of the model [24]. It was found that one of the components should have a very large value of H_1 to match the plastic modulus at the yielding. Another one had a smaller value of H_2 to satisfy the following $\sigma\sim\varepsilon$ relationship at or near the plastic strain limit, and the third component should have a very small value of C_3 to fit ε_r. This concept was basically the same as the concepts in the OW and OAK models, since by checking the determined parameters for the BS components in previously reviewed modeling approaches, it was found that some groups of H_i and C_i had very large values, which had a larger influence on simulated loop shapes, while some others had very small values, which controlled the simulated $\acute{\varepsilon}_r$. Therefore, it is possible to eliminate some BS components, leaving only three components as in [24] to simulate ratcheting behavior.

The above reviewed modeling approaches lead to the conclusion that the dynamic recovery terms play a vital role in the simulation of ratcheting of various materials. Abdel–Karim [66] reviewed a variety of modeling approaches and proposed several more equations of BSs, in which the only differences were found in the terms of dynamic recovery. In the current work, the simulation ability of several approaches is checked. If the already existing models are not suitable for simulating material responses of P91 at room temperature and 550 °C under multiple loading conditions, they should be further modified.

2.3 Advantage of Ferritic-Martensitic (FM) Steel

Ferritic–Martensitic steels include those ferritic steels with a martensite microstructure. The nominal compositions of a group of commercial and experimental FM steels are listed in Table 2.2.

Steel	C	S	Mn	Cr	Ni	Mo	W	V	Nb	B	N	Other
9Cr-1Mo (T9)	0.12	0.6	0.45	9	0.2	1						
Mod 9Cr-1Mo (T/P 91)	0.1	0.4	0.4	9	0.1	1		0.2	0.08		0.05	
E911	0.11	0.4	0.4	9	0.2	1	1	0.2	0.08		0.07	
NF616 (T/P 92)	0.07	0.06	0.45	9	0.25	0.5	1.8	0.2	0.05	0.004	0.06	
F82H	0.1	0.2	0.5	8			2	0.2		0.003		0.04 Ta
EUROFER	0.11	0.05	0.5	8.5			1	0.25		0.005		0.08 Ta
ORNL 9Cr-2WVTa	0.1	0.3	0.4	9			2	0.25			0.025	0.07 Ta
12Cr-1MoWV (HT9)	0.2	0.4	0.6	12	0.5	1	0.5	0.25				
HCM12A (T122)	0.11	0.1	0.6	12	0.3	0.4	2	0.25	0.05	0.003	0.06	1.0 Cu
CLAM	0.10	<0.002	0.45	9.0			1.5	0.2			<0.02	0.15 Ta
INRAFM	0.09-0.12	<0.002	0.4-0.6	8.8-9.2	<0.005	<0.002	1.3-1.5	0.2-0.24	<0.001	<0.001	0.02-0.04	0.06-0.08 Ta

Table 2.2: Nominal compositions of commercial and experimental FM steels [67-70]

Among the FM steels listed in Table 2.2, two of them belong to a more specific group, namely Grade 91 FM steels, including T9 and T/P91, where "9" indicates 9% Cr and "1" indicate 1% Mo. T9 is standard 9Cr–1Mo steel and T/P91 is modified 9Cr–1Mo steel. EUROFER belongs to the so-called reduced activation ferritic–martensitic (RAFM) steel because the 1% Mo is replaced by 1% W, which has a shorter half decay period. The other typical alloying elements, Nb, Ni, Cu, and N, also need to be eliminated or minimized in RAFM [67, 71, 72].

HT9 and T122 lie on the boundary between ferritic-martensitic steel and ferritic steel. The other FM steels in Table 2.2 have 8~9% Cr but the percentages of Mo and W are different from those of Grade 91 FM steel. CLAM is China Low Activation Martensitic steel [69] and INRAFM is Indian Reduced Activation Ferritic Martensitic steel [68, 70].
The material under investigation in the current work is P91, which belongs to mod. 9Cr–1Mo FM steel and further belongs to Grade 91 FM steel. Hence, more attention is paid to research on these types of steel.

Comparing to austenitic steels, such as types 316 and 304 stainless steel, FM steel has a lower thermal expansion coefficient and excellent irradiation resistance to void swelling [67, 73], owing to the non-compact crystal structure of ferrite. Void swelling limits the use of austenitic steels for fuel cladding and other in-core applications [67].

In high-temperature applications, Grade 91 FM steel can result in substantial reductions in component thickness compared to weaker alloys, such as Grade 22. The frequent startups, shutdowns, and load changes imposed by cycling duty lead to thermal fatigue, with the biggest challenges to the heat-recovery steam generator. By using higher-strength materials, such as Grade 91 FM steel, pressure-containing components can be made in thinner sections, which have smaller temperature gradient across the wall thickness and require less time to reach thermal equilibrium. This is an effective way to fight thermal fatigue [74].

The development of Grade 91 FM steel began in 1978 at Oak Ridge National Laboratories for the breeder reactor and was further developed by other researchers [75]. P/T-91 steels ("P91" for piping and "T91" for tubing) are modified Grade 91 steels with small additions of niobium (Nb), vanadium (V) and nitrogen (N) to give improved long-term creep properties [76].

An upgrade from the traditional P22 alloy to P91 can [74]:

- Reduce wall thickness by nearly two-thirds and component weight by 60%.
- Raise allowable strength in the 510~593 °C range by up to 150%.
- Raise the oxidation limit by 55 °C , enabling a lower corrosion allowance.
- Increase thermal-fatigue life by a factor of 10 to 12.

Modified Grade 91 FM steels are specifically intended for high-integrity structural service at elevated temperature, usually 500 °C or higher. These steels are now widely used for components such as headers, main steam piping, and turbine casings in fossil fueled power generating plants [76].

2.4 Microstructure of FM Steel

The superior properties of FM steel mentioned in the previous section have been attributed to a tempered martensitic microstructure consisting of dispersed carbide particles and a tangled dislocation substructure [77, 78]. As reported in [79] and [80] about mod. 9Cr–1Mo steel, the precipitates on prior austenite grain boundaries (PAGB) and lath boundaries were identified as $M_{23}C_6$ type carbides (M is Cr, Fe, Mo) whereas the precipitates inside the laths were identified as MX type carbides (M was V, Nb and X was carbon and nitrogen).

The excellent properties of FM steel depend entirely on the creation of a precise microstructure by heat treatment, and on the preservation of this microstructure throughout its service life. Failure to obtain this precise microstructure in production can seriously degrade the alloy's high-temperature properties. This is different from traditional carbon and low-alloy steels such as Grade 11 and 22 (operating at the low stresses typical of power applications), which are less sensitive to microstructure change [74].

Heat treatment of the Cr-Mo and Cr-W FM steels is crucial to induce the required microstructure [74]. These steels are firstly normalized, then air-cooled, and tempered afterwards [67]. Detailed heat treatment of, for example, mod. 9Cr–1Mo steel is as follows: the alloy is firstly heated above its upper critical transformation temperature (AC_3 line) for 0.5~1 hour until it is fully austenitic. Then, the steel is cooled in air below 200 °C for the full transformation of austenite into untempered martensite, which is very strong but brittle [74, 77]. The material is then tempered at around 760 °C to improve ductility and toughness and to induce the formation of critical carbide and carbo-nitride precipitates [79]. However, over-tempered mod. 9Cr–1Mo steel can have a substantially higher creep rate at temperature of 560 °C and a much lower hardness value (<180 on the Vickers Hardness scale or HV, instead of the expected 200+ HV). In addition to incorrect heat treatment, any action that alters the precise microstructure of the steel, such as hot bending, forging, and welding which regularly occurs during component fabrication and plant construction, can lead to failure to achieve superior high-temperature properties [74].

On the other hand, microstructure stability during the whole service life has the same importance as the precise microstructure achievement during production. The fracture toughness of many power plant steels deteriorates during service at elevated temperatures owing to evolution of carbides and intermetallic phases and segregation of tramp elements (e.g. P, As, Sn) to PAGB [80]. It was also reported in [81] that standard 9Cr–1Mo steel was susceptible to temper embrittlement.

The differences between microstructures of standard and modified 9Cr–1Mo steel are discussed next: The sequence of carbide precipitation processes in mod. 9Cr–1Mo steel is consistent with those for the standard composition. However, the replacement of Cr_2C in the modified alloy during tempering by arrays of fine vanadium carbide particles along the lath interfaces leads to a significant improvement in microstructural stability at temperature up to 650 °C, even under static tensile and creep conditions. As a result, the lath morphology in the modified alloy remains intact for long periods at temperatures up to 650 °C owing to the interfacial pinning by vanadium carbide precipitates, which coarsen very slowly [77]. The average prior austenite grain size (PAGS) of modified alloy is 20μm, which is smaller than that of standard alloy (~40μm). This is attributed to the presence of un-dissolved carbides along the austenite grain boundaries during normalization treatment, which inhibits the growth of austenite grains [80].

It was reported that the lath structure was retained, at least in certain regions, even after 10000 h of aging at elevated temperatures. Although carbides in mod. 9Cr–1Mo steel grew with aging time and temperature, the coarsening of V(Nb) carbides was negligible compared to $M_{23}C_6$ carbides [80]. In the current work, all experiments were done either at room temperature or supposedly for no more than 100 hours at 550 °C, which is a much shorter time than that of aging tests (e.g., >5000 hours reported in [80]). Hence, the change of microstructure owing to aging is negligible for experiments in the current work.

Another concern with FM steel is the oxidation during experiments at elevated temperatures. However, according to [82-84], oxidation occurs only at the surface. It was reported that the oxide layer would assist crack initiation and propagation for specimens tested under compressive hold [82]. A detailed investigation was carried out on the oxidation of mod. 9Cr–1Mo steel at 550 °C: The thickest oxide layer on the surface was found to be 20 μm for a high strain range creep-fatigue experiment with a fatigue strain range of 0.7% and a creep range 0.5%. In other experiments, such as pure fatigue and low strain range creep-fatigue tests, the oxide layers were only 2–3 μm thick. In static oxidation tests, an extreme case was that the oxide thickness was merely 2.80 μm after 36 days of oxidation at 550 °C [84]. Since the specimens in the current work are 8.8 mm in diameter, which is much larger than the possible thickness of oxide layer, the oxidation is not taken into account.

The microstructural evolution of FM steel during ratcheting is rarely reported in the literature. The only report was about the microstructure of steel X12CrMoWVNbN10-1-1 after stress-controlled creep-fatigue loadings at 600 °C. The collapse of martensitic laths after the ratcheting tests was observed. It was found that such collapse of laths gradually disappeared with decreasing grade of unloading [27, 85].

Such collapse of laths was not reported for FM steel in creep tests or without loading [77, 80, 86-88]. However, in strain-controlled LCF tests, a similar disappearance of martensitic laths was observed in high-chromium martensitic GX12CrMoVNbN9-1 (GP91) cast steel at RT, 550 °C, and 600 °C. Such disappearance was suggested to be the dominant factor in the acceleration of fatigue softening of the material [89]. Another observation was on mod. 9Cr–1Mo steel in LCF tests, which mentioned the conversion of the initial heavily dislocated lath structure to equiaxed cells with low dislocation density and coarse carbides. This conversion was suggested to cause cyclic softening [82, 83, 90]. A detailed review of work on cyclic softening is given in the next section.

Therefore, according to these reports ([27, 82, 83, 85, 89, 90]) the collapse/disappearance/conversion of the martensitic lath structure is common under cyclic loading, either stress- or strain-controlled.

2.5 Cyclic Softening of FM Steel

As mentioned in the previous sections, one of the main disadvantages of FM steel is cyclic softening, which was reported for various FM steels, such as 9Cr–1Mo steel [24, 73, 91-95], EUROFER [1, 96], F82H [97, 98], P92 [93], and HT-9 [99]. The reported causes of cyclic softening are decrease of dislocation density by cell structure formation [73], conversion of lath structure to equiaxed cells [82, 83, 90, 92, 99], disappearance of lath [89], annihilation of low angle boundary [91], coarsening of laths and subgrains [73, 94], and coarsening of precipitate [99].

In one of the earliest reports on cyclic softening of FM steel, the cause of cyclic softening was cited as rearrangement of dislocations previously introduced by the quenching. LCF tests were carried out on ferrite-pearlite steel AISI 420, FM steel MANET II, and RAFM steel F82H mod. No cyclic softening occurred on AISI 420 (RT–550 °C), but obvious cyclic softening occurred on MANET II (150–550 °C) and F82H mod. (550 and 650 °C) [100].

After comparing the LCF behaviors of P91 and P92 steels, it was found that the softening rate of P92 steel increased with increase in strain amplitude whereas the softening rate of P91 remained constant with strain amplitude [93]. This indicates the modeling of cyclic softening for P91 can be suitable for a large range of strain amplitudes (ε_a = 0.25–0.60% in [93]).

A typical LCF test on mod. 9Cr–1Mo FM steel is illustrated in a stress-strain diagram in [24], as shown in Fig. 2.10: The strain range kept constant but both peak tensile stress and peak compressive stress decreased as the number of cycles increased. It was suggested that the cyclic softening was the reason for the so-called "progressive deformation instability" in ratcheting tests [24], as shown in Fig. 2.3 and Fig. 2.4 in Section 2.1.

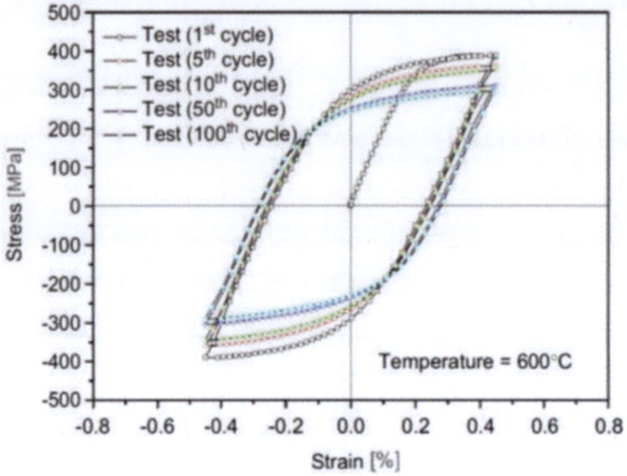

Fig. 2.10: Test results of the cyclic softening characteristics of mod. 9Cr–1Mo steel at 600 °C (strain-controlled) [24].

As mentioned in Section 2.2, the cyclic softening was taken into account in the simulation of the ratcheting behavior of the mod. 9Cr–1Mo steel at 550 °C in the YT model [7]. The cyclic softening behavior was expressed through variation of the asymptotic values of parameter r_i, which was given as:

$$r_i = r_i^0 - r_i^s \tag{2-23}$$

$$\dot{r}_i^s = d(\bar{r}_i^s - r_i^s)^q \dot{p} \tag{2-24}$$

$$\dot{\bar{r}}_i^s = \langle r_0 - \bar{r}_i^s \rangle \tag{2-25}$$

where r_i^0, d, q and h_i are material and temperature-dependent parameters. The value of H_i decreased from the initial value of r_i^0 by subtracting r_i^s which expressed the progress of the cyclic softening.

By recalculation using the YT model, the result is shown in Fig. 2.11:

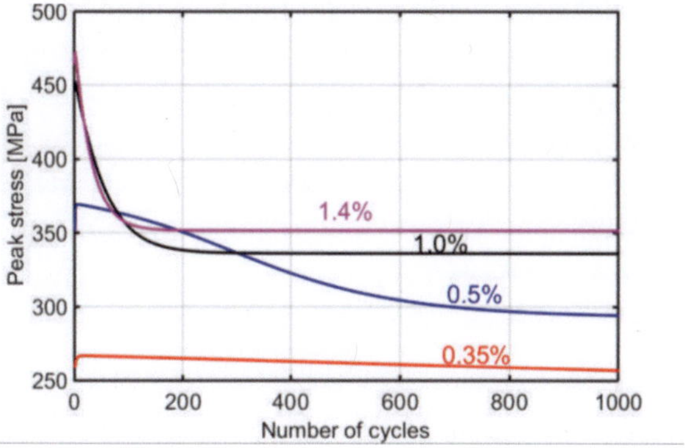

Fig. 2.11: Peak stress vs. number of cycles in Yaguchi–Takahashi model description for strain-controlled LCF tests performed with various strain amplitudes

It is clear that the simulated σ_{peak} decreases with increasing cycle number and it reaches a saturation stage with negligible decrease. This is, however, different from the experimental results, such as those reported in [1], in which a saturation stage (stage 2) was reached after the initial fast decrease of σ_{peak} (stage 1) but σ_{peak} still decreased linearly with a constant slope until a macro crack appeared.

Another approach is to introduce an isotropic hardening factor R into eq. (2-1) of the Chaboche model, as in eq. (2-26) [101]:

$$\dot{\boldsymbol{\varepsilon}}^{in} = \frac{3}{2}\langle\frac{\Sigma_{eq}-K-R}{Z}\rangle^{n}\frac{\boldsymbol{\Sigma}}{\Sigma_{eq}} \quad (2\text{-}26)$$

in which R is the key to expressing the cyclic softening. The equation for R is as follows:

$$R = hp + Q(1 - e^{-bp}) \tag{2-27}$$

in which h, Q, and b are material- and temperature-dependent parameters. The term hp is the linear part of eq. (2-27), which simulates stage 2 of the cyclic softening. As shown in Fig. 2.12, h indicates the slope of stage 2. The second term in eq. (2-27) represents stage 1 of the cyclic softening. Q was estimated as the difference between point X and σ_{peak} in the first cycle in Fig. 2.12 while parameter b was the speed to reach σ_{peak} at the end of stage 1 [101].

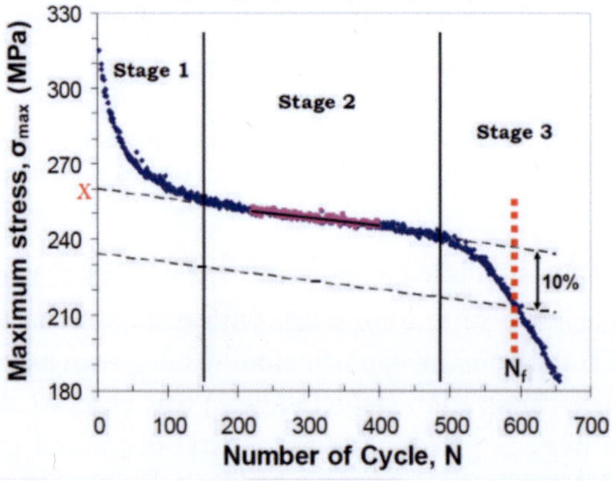

Fig. 2.12: The results of P91 strain-controlled tests at 600°C [101].

Note that Fig. 2.12 illustrates how to define the number of cycle to failure, namely N_f. It is defined according to BS7270:2006 standard as the cycle during which σ_{peak} has decreased by 10% from that predicted by extrapolation of the saturation curve (stage 2) [101].

The cyclic softening of RAFM steel, including F82H mod and EUROFER97, was evaluated with LCF tests and a constitutive model was built correspondingly [1]. This model proposed in [1] is referred to as the Aktaa–Schmitt (AS) model. In the current work, the influence of cyclic softening on ratcheting behavior is evaluated and the AS model is applied for the cyclic softening, since both EUROFER97 and P91 are Grade 91 FM steels.

The AS model follows the way of the Chaboche model with the AF rule as in eqs. (2-1)–(2-6). Eq. (2-2) was modified with a softening factor ψ to describe cyclic softening (see eq. (2-28)). In eq. (2-31), the last term represents the static recovery of cyclic softening. Further, the AF rule (see eq. (2-6)) was applied with static recovery of kinematic hardening in the Aktaa–Schmitt model:

$$\boldsymbol{\Sigma} = \frac{\boldsymbol{s}}{\psi} - \boldsymbol{\Omega}_i \tag{2-28}$$

$$\begin{gathered} \psi = \psi_1 + \psi_2 \\ with\ \psi_1(t=0) = 0\ and\ \psi_2(t=0) = 1 \end{gathered} \tag{2-29}$$

$$\dot{\psi}_1 = -h\dot{p} \tag{2-30}$$

$$\dot{\psi}_2 = c(\psi_s - \psi_2)\dot{p} - r_\psi |\psi_2 - \psi_r|^{m_\psi - 1}(\psi_2 - \psi_r) \tag{2-31}$$

$$\psi_s = 1 - \psi_{s,\infty}(1 - exp(-c_s \max_{-\infty<\tau<t} |\varepsilon_{eq}^{in}(\tau)|)) \tag{2-32}$$

$$\varepsilon_{eq}^{in}(\tau) = \sqrt{\frac{2}{3}\varepsilon^{in}(\tau):\varepsilon^{in}(\tau)} \tag{2-33}$$

The AS model was used to successfully simulate strain-controlled LCF tests on EUROFER97 at 450 and 550 °C, as well as tests on F82H mod at 450, 550 and 650 °C [1]. Note that the AS model assumes only one BS, instead of four, eight, or even 12 BS components used in other modeling approaches [3, 4, 6, 30]. Therefore, one BS is enough for modeling of material responses under strain-controlled loadings.

3 Material and Approach

The current work is organized according to previous research reviewed in Chapter 1, especially on mod. 9Cr–1Mo FM steels. The task is to investigate the ratcheting behavior of mod. 9Cr–1Mo steel (P91) at RT and 550 °C.

Since no study on the ratcheting of mod. 9Cr–1Mo FM steel at RT has been reported, this shall be carried out in the current work. On the other hand, although there have been reports such as [7, 24-28] about the ratcheting of FM steel at high temperatures, only Yaguchi and Takahashi [7, 26] have reported about the ratcheting at 550 °C. As mentioned in Section 2.1, some tests reported in [26] were still doubtful, since ratcheting tests performed with ε_{peak} = 400 MPa contradict the ultimate tensile strength of 374 MPa of T91 reported in SCK•CEN [55], since both materials mentioned in [26] and [55] are mod. 9Cr–1Mo FM steel.

The tension-compression asymmetry reported in [7, 26] should also be double checked, since no other reports [1, 3, 4, 6, 24-28, 74, 82, 102] mentioned such an asymmetry of FM steel at high temperature. Furthermore, in the modeling approach in [7], the tension-compression asymmetry was expressed by a term of hydrostatic pressure to modify the yield stress, which is also in question, because steel is supposed to be a solid material hence no influence of hydrostatic stress exists, although there were debates about the influence of hydrostatic pressure, such as in Jung [103], Casey and Sullivan [104], and Drucker [105].

Before stress-controlled ratcheting tests, strain-controlled LCF tests are performed to evaluate the cyclic softening of the material because softening can accelerate ratcheting.
The process of the work in the current research is illustrated in Fig. 3.1.

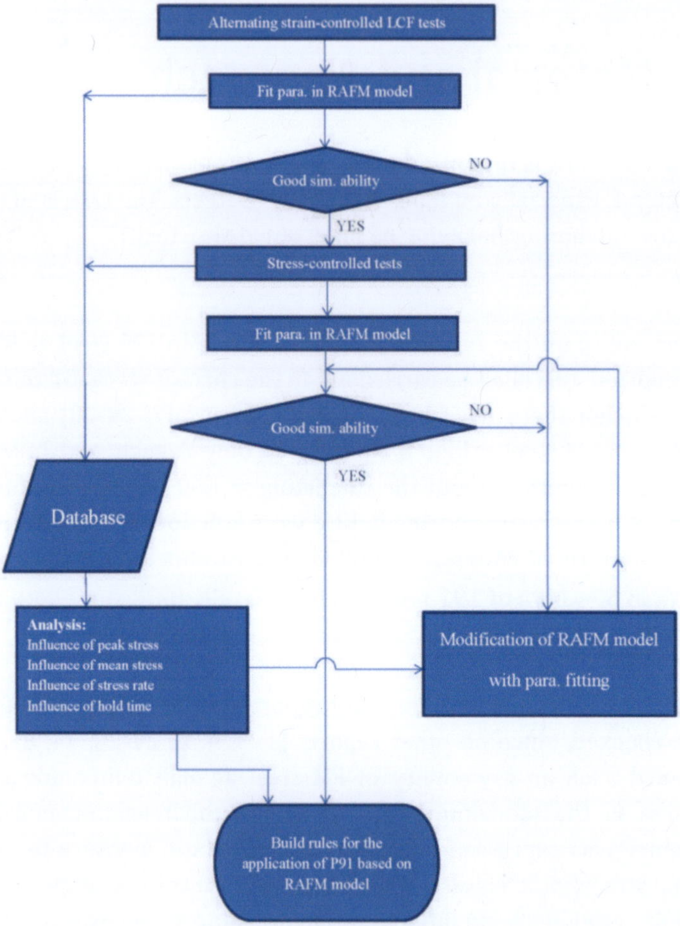

Fig. 3.1: Process of research on ratcheting behavior of P91.

3.1 Specimens and Facilities

The steel plate of P91 was provided by French Alternative Energies and Atomic Energy Commission (CEA) with the specification RM2432 of the

RCC-MR code (1993 edition). The plate was austenitized at 1050 °C for 30 min, quenched, and then tempered at 780 °C for 1h. Both austenitizing and tempering temperatures are 10~20 °C higher than those reported in [74] and [77].

The chemical compositions are listed in Table 3.1

Elements	% Mass
C	0.086
Cr	8.910
Mo	0.917
Mn	0.363
Si	0.324
V	0.198
Ni	0.149
Nb	0.08
Cu	0.068
N_2	0.041
Al	0.018
P	0.017
As	0.012
W	0.01
Sn	0.005
Ti	0.002
O_2	0.002
S	0.001
B	0.001
Zr	0.001

Table 3.1: Chemical compositions of P91

The specimens were fabricated from a steel plate with a thickness of 30 mm. The fabrication was carried out in three steps. First, raw samples with diameter of 16 mm and length of 80 mm were cut from one plate by wire EDM (electrical discharge machining) technique. Afterwards, the specimens were produced by turning to an entire length of 77 mm (gauge length is 23 mm) with a nominal diameter of 8.8 mm. Finally, the surface

was radially polished to remove the turning grooves. Fig. 3.2 shows a technical drawing of the P91 specimens with a photo of a real one.
The experiments were performed on a SCHENCK hydraulic testing machine controlled by Instron 8800. The controlling programs were written in DASYLab with graphic programming language. Strain/stress ranges and rates were used as input test parameters for the DASYLab program.

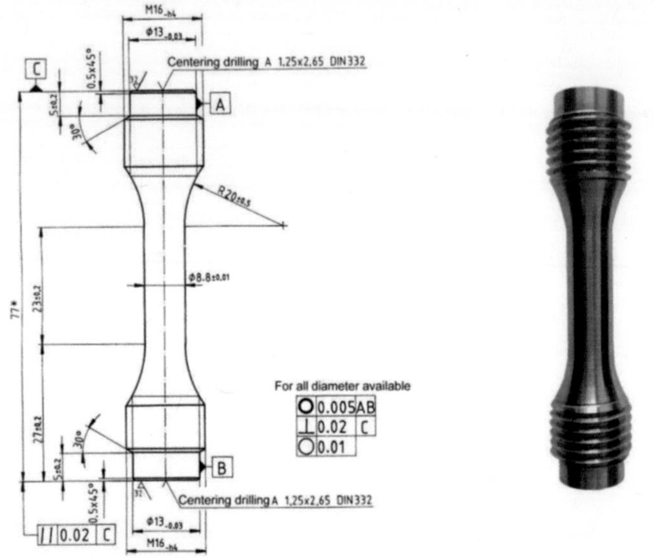

Fig. 3.2: Technical drawing of P91 specimen together with a real photo.

3.2 Experimental Planning for Testing at RT

3.2.1 Strain-controlled Tests at RT

A tensile test with strain rate 1×10^{-4}/s was performed at the very beginning. The material at RT has an ultimate tensile strength 666.8 MPa with a uniform elongation of 9.7%. Yield strength with an offset of 0.2% was found to be 481.8 MPa.

There are two stages of experiments at the same temperature. The first stage includes alternating strain-controlled LCF tests to evaluate the cyclic softening of P91. As listed in Table 3.2, the strain range is 0.6–1.5% with a strain rate of $\pm 3 \times 10^{-4}$/s. Data points are recorded every 0.02 second, hence at least 2000 data points are sampled in one cycle, which is quite enough to represent the material responses. The results of the strain-controlled tests at RT are presented in Section 4.1.

Strain range $\Delta\varepsilon$ (%)	Strain rate
0.6	
0.8	
1.0	$\pm 3\times10^{-4}$/s
1.2	
1.5	

Table 3.2: Symmetric alternating strain-controlled LCF tests at RT

3.2.2 Stress-controlled Tests at RT

After the strain-controlled LCF tests, the Young's modulus and yield strength were determined, with which the stress ranges in the following stress-controlled tests were roughly determined. The chosen stress should at least induce a measurable inelastic strain since no ratcheting exists with only elastic deformation. On the other hand, the imposed stress should be not too high. At RT, the peak tensile stress applied on the specimen was limited to 550 MPa where the corresponding total strain is around 1%.

Owing to a large difference between σ_{eng} and σ_{true} with high deformation, the recorded data mix both material response and influence of structure. For instance, when engineering stress σ_{eng} = 550 MPa and ε = 5%, the corresponding true stress σ_{true} = 577.5 MPa. Therefore, the real time deformation is taken to modify the signal, which controls the loading on specimen, to adapt the reduced cross section of specimen with high strain. In the following analytical modeling process, pure material responses are required to compare with model description. Hence, all stress-controlled tests listed in Table 3.3 are true-stress-controlled, instead of engineering-stress-controlled.

The schema for true-stress controlling is illustrated in Fig. 3.3. In the controlling software DASYLab, there is a module called "signal generator" that can generate a signal in triangle wave. The real-time deformation is measured by the extensometer and transformed to the real-time total strain through the module of the data processor. When the signal generator is supposed to generate a loading of, for example, 500 MPa and the real time total strain is ε = 1%, the output of the signal generator is 500/1.01 = 495 MPa. Consequently, the current engineering stress is σ_{eng} = 495 MPa, but σ_{true} = 500 MPa, according to eq. (3-1):

$$\sigma_{true} = \sigma_{eng}(\varepsilon + 1) \tag{3-1}$$

The true-stress controlling is based on the von Mises criterion with no volumetric plastic strain.

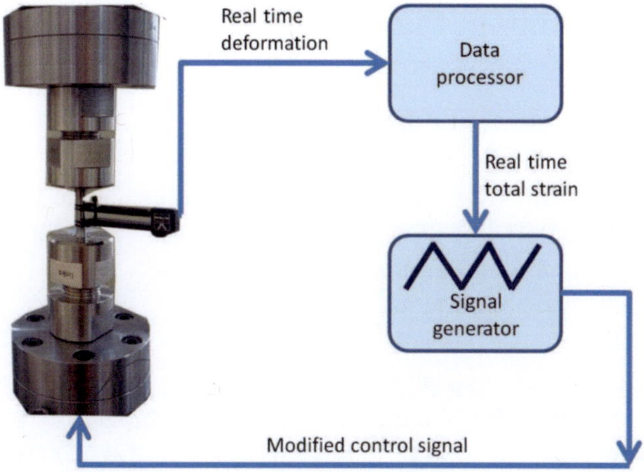

Fig. 3.3: Schema of true stress controlling

Several groups of experiments are planned to explore the influences of main factors to ratcheting, as listed in Table 3.3. The results of the stress-controlled tests at RT are presented in Section 4.2.

a) Influence of peak stress

σ_{max} [MPa]	550	520	500	480	460	450	430	400
σ_{min} [MPa]	-500	-470	-450	-430	-410	-400	-380	-350
Stress ratio	-0.909	-0.904	-0.9	-0.896	-0.891	-0.889	-0.884	-0.875
σ_{mean} [MPa]	25							
$\dot{\sigma}$ [MPa/s]	±50							

b) Influence of mean stress/stress ratio

σ_{max} [MPa]	500										
σ_{min} [MPa]	0	-300	-400	-430	-440	-450	-460	-480	-500	-510	-550
Stress ratio	0.00	-0.6	-0.8	-0.86	-0.88	-0.9	-0.92	-0.96	-1.0	-1.02	-1.1
σ_{mean} [MPa]	250	100	50	35	30	25	20	10	0	-5	-25
$\dot{\sigma}$ [MPa/s]	±50										

c) Ratcheting with zero mean stress

σ_{max} [MPa]	550	520	500	480	450
σ_{min} [MPa]	-550	-520	-500	-480	-450
Stress ratio	-1				
σ_{mean} [MPa]	0				
$\dot{\sigma}$ [MPa/s]	±50				

d) Influence of stress rate

σ_{max} [MPa]	500		
σ_{min} [MPa]	-450		
$\dot{\sigma}$ [MPa/s]	±10	±50	±250

e) Influence of hold time

σ_{max} [MPa]	500	
σ_{min} [MPa]	-450	
$\dot{\sigma}$ [MPa/s]	±50	
Hold 10 min at peak	Tension	Compression

Table 3.3: Stress-controlled uniaxial tests at RT

3.3 Experiment Planning for Testing at 550 °C

3.3.1 Strain-controlled Tests at 550 °C

A tensile test with strain rate 1×10^{-4}/s was performed at the very beginning. The material at 550 °C was found to have an ultimate tensile strength 386.1 MPa with uniform elongation of 1.6%. Yield strength with an offset of 0.2% is 353.7 MPa.

At 550 °C, similar uniaxial symmetric strain-controlled LCF tests are carried out on P91. The five tests with total strain ranging $\Delta\varepsilon$ = 0.6–1.5% are listed in Table 3.4. The strain rates are $\pm 3 \times 10^{-4}$/s and the sampling interval is 0.02 seconds. The results of the strain-controlled tests at 550 °C are presented in Section 5.1.

Strain range $\Delta\varepsilon$ (%)	Strain rate
0.6	$\pm 3 \times 10^{-4}$/s
0.8	
1.0	
1.2	
1.5	

Table 3.4: Symmetric alternating strain-controlled LCF tests at 550 °C

3.3.1 Stress-controlled Tests at 550 °C

The high-temperature stress-controlled tests on P91 are essentially the same as those at RT; the only difference is the lower imposed stresses owing to reduced strength at high temperature. The maximum stress in the following cyclic tests is limited to 350 MPa, which is approximately equal the yield strength at 0.2% offset at 550 °C.

The planned ratcheting tests at 550 °C are listed in Table 3.5. Creep tests are performed (see Table 3.6) to compare with the ratcheting tests with hold times (see Table 3.5e). The results of stress-controlled tests at 550 °C are presented in Section 5.2.

Note that the high-temperature stress-controlled tests are also true-stress-controlled, as with the RT tests, by modifying the output signal with real-time deformation, as discussed in Section 3.2.2.

a) Influence of peak stress

σ_{max} [MPa]	350	335	315	300	275
σ_{min} [MPa]	-300	-285	-265	-250	-225
Stress ratio	-0.86	-0.85	-0.84	-0.83	-0.82
σ_{mean} [MPa]	25				
$\dot{\sigma}$ [MPa/s]	±50				

d) Influence of stress rate

σ_{max} [MPa]	325		
σ_{min} [MPa]	-310		
Stress ratio	-0.95		
σ_{mean} [MPa]	7.5		
$\dot{\sigma}$ [MPa/s]	±10	±50	±250

b) Influence of mean stress/stress ratio

σ_{max} [MPa]	325	325	325	325	325	325	325	325	325	325	325	325	325	325	325	325
σ_{min} [MPa]	0	-125	-175	-225	-250	-260	-275	-285	-295	-300	-305	-310	-315	-320	-330	-335
Stress ratio	0	-0.39	-0.54	-0.69	-0.77	-0.80	-0.85	-0.88	-0.91	-0.92	-0.94	-0.95	-0.97	-0.98	-1.02	-1.03
σ_{mean} [MPa]	162.5	100	75	50	37.5	32.5	25	15	15	12.5	10	7.5	5	2.5	-2.5	-5
$\dot{\sigma}$ [MPa/s]	±50															

c) Ratcheting with zero mean stress

σ_{max} [MPa]	350	340	325	315
σ_{min} [MPa]	-350	-340	-325	-315
Stress ratio	-1	-1	-1	-1
σ_{mean} [MPa]	0			
$\dot{\sigma}$ [MPa/s]	±50			

e) Influence of hold time

σ_{max} [MPa]	325		
σ_{min} [MPa]	-310		
Stress ratio	-0.95		
σ_{mean} [MPa]	7.5		
$\dot{\sigma}$ [MPa/s]	±50		
Hold time in Ten.[min]	5	0.5	0
Hold time in Comp.[min]	0	0	5

Table 3.5: Stress-controlled uniaxial tests at 550 °C

Stress (MPa)	325	325	-310	275	200
Control type	Engineering-stress controlled	True-stress controlled	True-stress controlled	True-stress controlled	True-stress controlled

Table 3.6: Creep tests at 550 °C

4 Experiments at Room Temperature

In this chapter, the results of the experiments performed at room temperature are reported. In Section 4.1, the experimental results from the strain-controlled LCF tests are illustrated. In Section 4.2, the results of the stress-controlled tests are presented in sub-sections according to various influencing factors.

4.1 Results of Strain-controlled Tests

In the first stage of experiments, alternating strain-controlled LCF tests were carried out. Fig. 4.1 illustrates one of the tests performed with a strain range of $\Delta\varepsilon$ = 1.5%. Since the strain is in a triangle wave, the magnitude of the strain rate is constant, as shown in Fig. 4.1a. The stress is yielded as a consequence (see Fig. 4.1b). The strain rate is chosen to be $\pm 3 \times 10^{-4}/s$, hence the yielded stress has a stress rate of around ± 50 MPa/s in the elastic region, compared to ± 50 MPa/s as the chosen $\dot{\sigma}$ in the following stress-controlled tests in the second stage of the RT experiments.

The experiment was ended after a macro crack was observed with which a kink on the compressive half of hysteresis loop appears, as shown in Fig. 4.1c. The appearance of this kink is owing to closing of a macro crack: During one cycle of alternating stress, the crack is opened gradually under tension and joins together rapidly under compression, and consequently a kink appears on the hysteresis loop. In strain-controlled LCF tests at RT, the measured peak stresses do not smoothly decrease in every test, instead, they behave in a random way. The reason is that the specimen section with homogenous radius has a length of 23 mm, but the gage length of the extensometer is only 10 mm. Therefore, the crack does not always appear within the gage length of the extensometer. Hence, the reliable data in the LCF test at RT is only taken until a roughly defined cycle N_0 within the stage where the peak stress is linearly decreasing (saturation stage of cyclic softening) and near the cycle where the deviation of peak stress from the linear decrease appears due to the macro crack. Further, the lifetime is roughly determined as the total cycle number, N_0.

Fig. 4.1d shows the hysteresis loops in three cycles. The peak stresses for the first and 300th cycles are obviously different while the difference between the 300th and 800th cycles is relatively small. The change of σ_{peak} versus number of cycles is illustrated in Fig. 4.1e and f.

It was found that σ_{peak} increases in the initial 5–10 cycles owing to cyclic hardening. As shown in Fig. 4.1e, the σ_{peak} reaches its maximum in the sixth cycle in this test performed with $\Delta\varepsilon$ = 1.5%. After this initial increase, σ_{peak} decreases (see Fig. 4.1f) which shows the obvious cyclic softening of P91. In the saturation stage of cyclic softening, the peak stress is still decreasing, although it is not obvious in the view of Fig. 4.1f. The σ_{peak} after around the 1000th cycle has an accelerated decrease owing to macroscopic crack propagation.

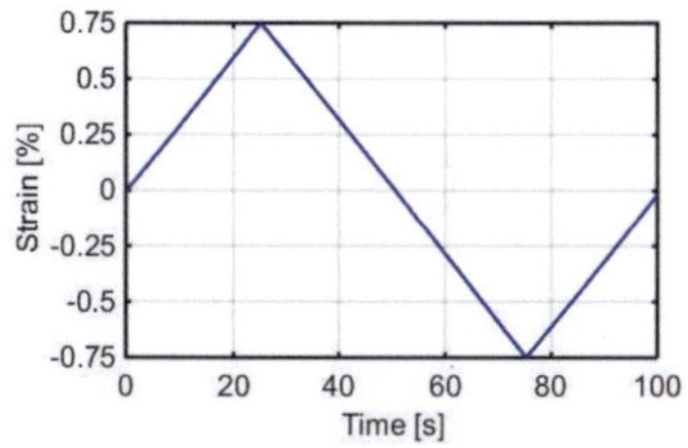

a) *1st cycle, strain vs. time.*

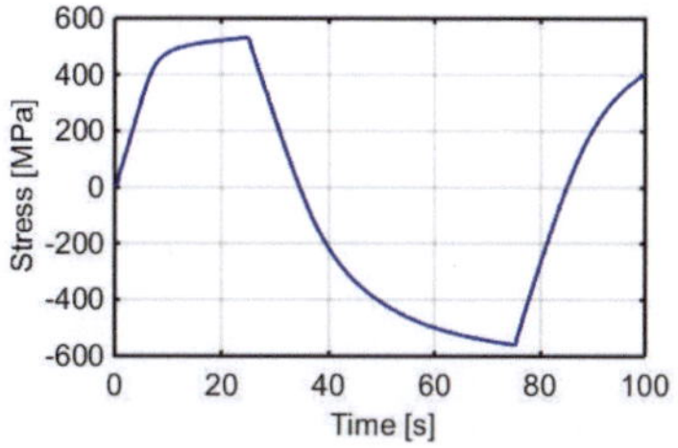

b) *1st cycle, stress vs. time.*

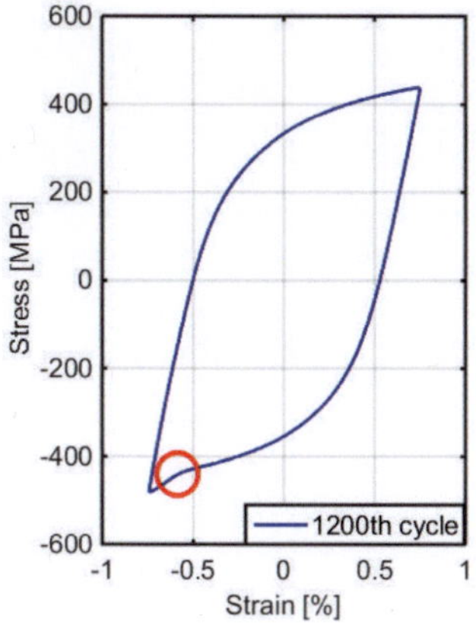

c) *Hysteresis loops of 1200th cycle*

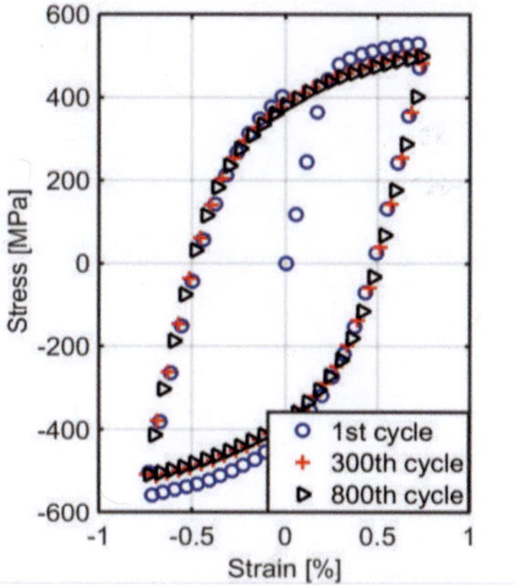

d) *Hysteresis loops of 1st, 300th, 800th cycle.*

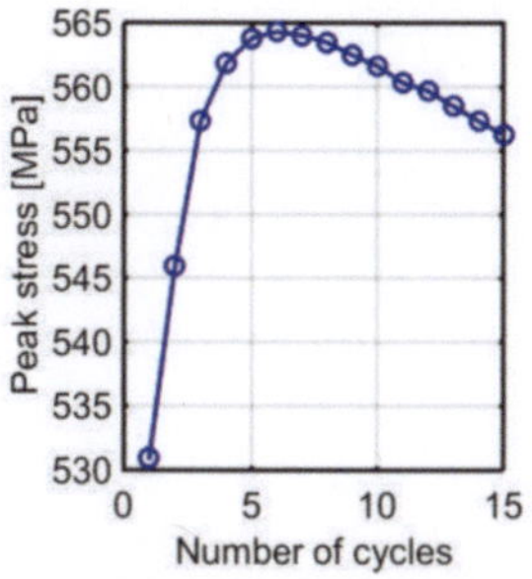

e) *σ_{peak} in initial 15 cycles.*

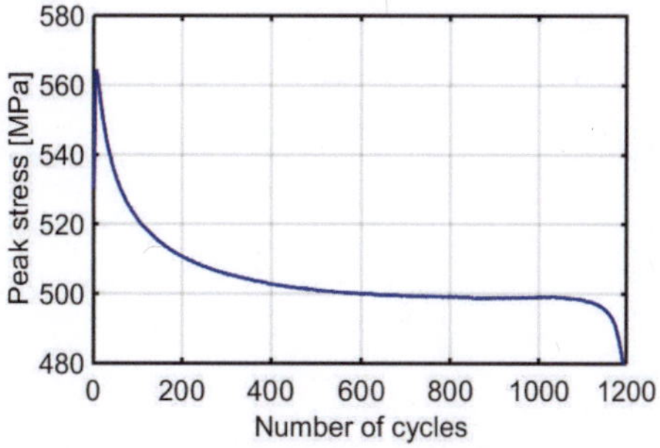

f) *σ_{peak} in cycles till fracture*

Fig. 4.1: Symmetric strain-controlled LCF tests at RT, $\Delta\varepsilon$=1.5%.

Cyclic softening was also found in the other tests with $\Delta\varepsilon$ = 0.6%, 0.8%, 1.0%, and 1.2%, as illustrated in Fig. 4.2c. According to the curves of peak tensile stresses versus number of cycles, the cyclic softening is more rapid in the initial stage. After the initial fast softening, σ_{peak} has a linear and slower decrease. The second stage is also called the saturation stage, as reviewed in Section 2.5. This stage lasts until the appearance of a macro crack.

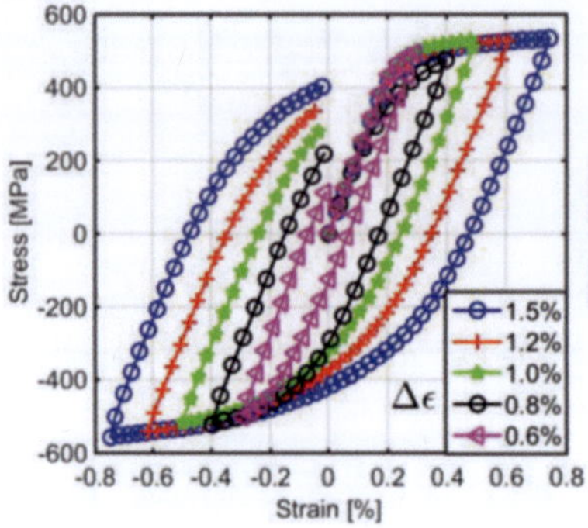

a) *Hysteresis loops of the first cycle.*

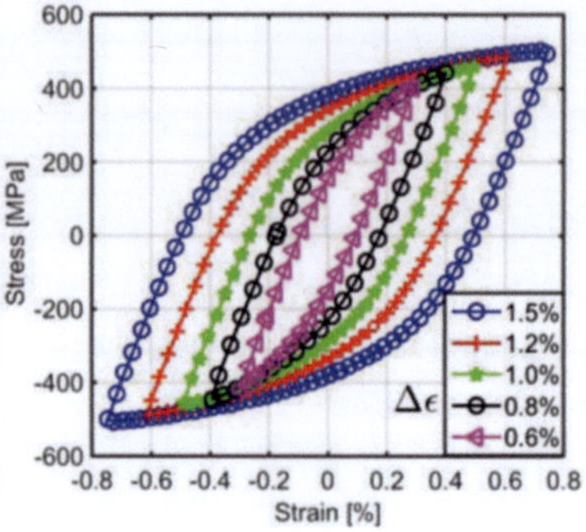

b) *Hysteresis loops of the cycle at* $N_0/2$

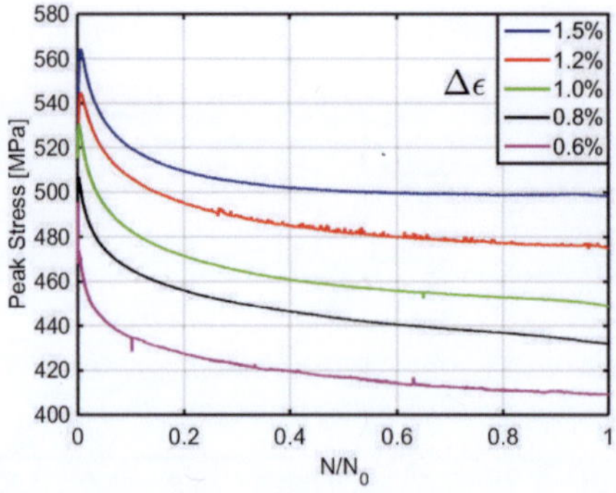

c) *Peak tensile stresses vs. normalized number of cycles*

Fig. 4.2: Strain-controlled LCF tests at RT performed with strain ranges 0.6% ~ 1.5%.

As with the peak tensile stresses, the magnitudes of the peak compressive stresses decrease during the LCF test. However, an interesting phenomenon was found in these strain-controlled tests: The magnitude of peak compressive stress is larger than the corresponding peak tensile stress in every cycle, as shown in Fig. 4.3. In these diagrams, red curves indicate the absolute value of the peak compressive stresses and the blue curves indicate the peak tensile stresses. The larger strain range leads to a larger difference between the magnitudes of the peak tensile and compressive stresses. In the test performed with $\Delta\varepsilon = 0.6\%$, this difference in the saturation stage was around 6 MPa while with $\Delta\varepsilon = 1.5\%$, this difference was around 12 MPa.

One reason for this difference is the difference between σ_{eng} and σ_{true}. It is well known that σ_{true} is larger than σ_{eng} under tension owing to reduction of cross section area (the other way around under compression). σ_{eng} shown in Fig. 4.3 is transferred into σ_{true} shown in Fig. 4.4. Although the differences between tensile and compressive peaks are smaller in true-stress cases, they still cannot be ignored: for example, around 3.5 MPa for $\Delta\varepsilon = 0.6\%$ and around 4 MPa for $\Delta\varepsilon = 1.5\%$.

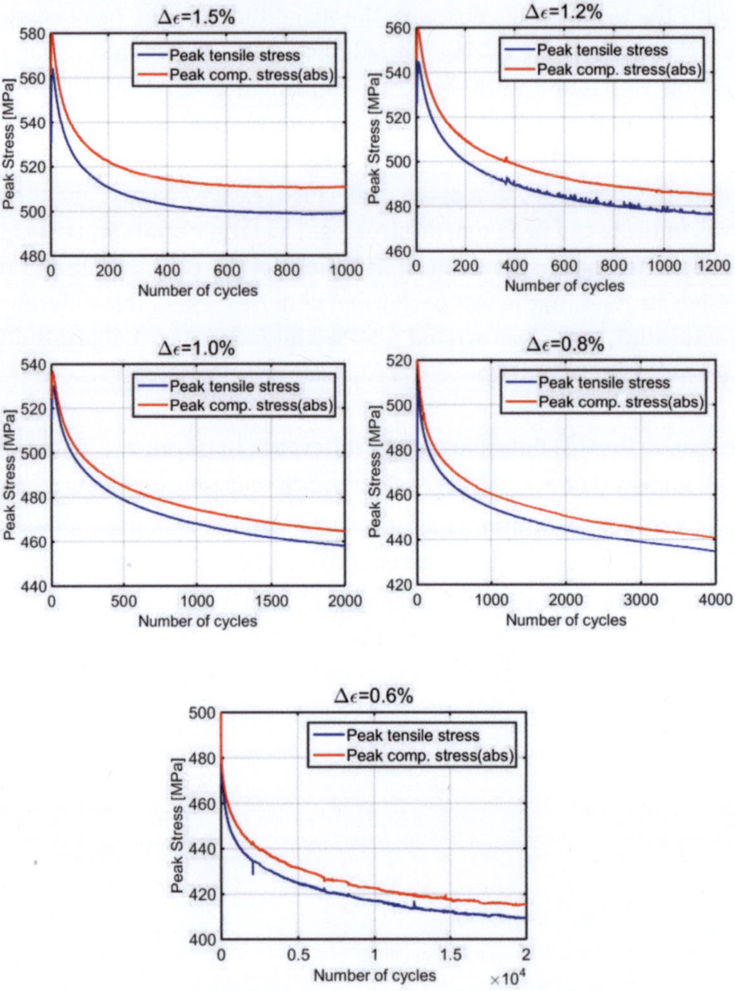

Fig. 4.3: Engineering peak compressive stresses (abs.) compared with engineering peak tensile stresses, at RT.

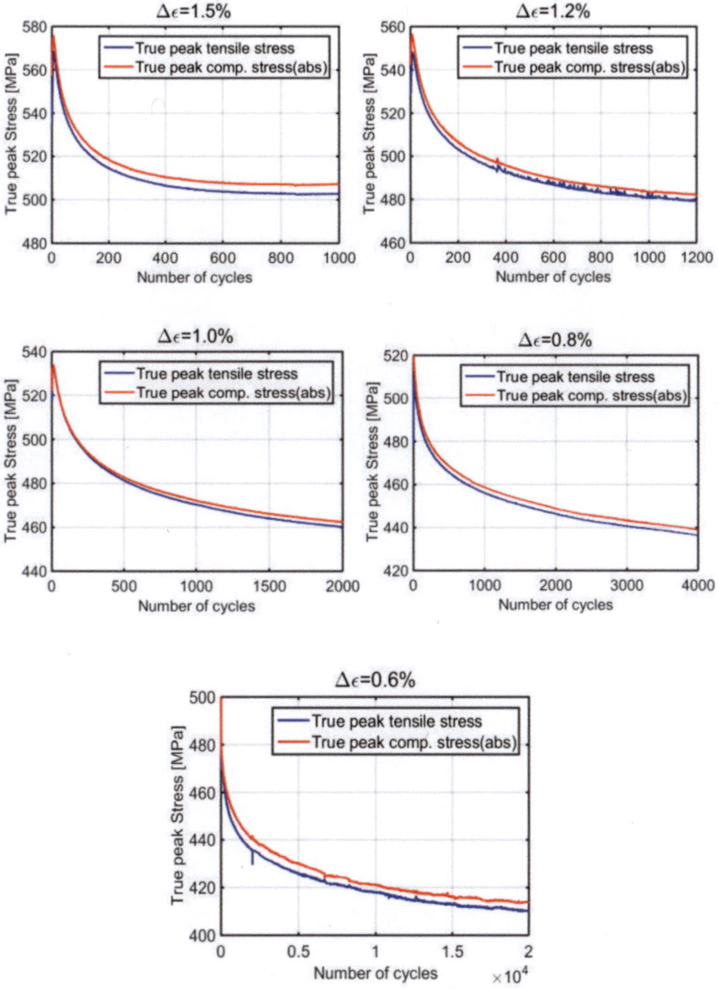

Fig. 4.4: True peak compressive stresses (abs.) compared with true peak tensile stresses, at RT.

According to the analysis above, such asymmetry of peak tensile and compressive stresses is obviously material behavior, which indicates that the material strength under tension and compression is not the same at RT. This asymmetry can yield positive ratcheting in stress-controlled tests with zero σ_{mean} in the second stage of RT experiments. The reason is that the symmetry of the strain leads to the asymmetry of the stress, hence in the other way around, the symmetry of the stress should lead to the asymmetry of the strain. Since peak compressive stresses are larger than peak tensile stresses in symmetric strain-controlled LCF tests, when the stress become symmetric, the change of strain during stress-increasing process should be larger than that during the stress-decreasing process in the symmetric stress-controlled tests. As a result, positive ratcheting appears with zero σ_{mean}. As reviewed in Section 2.1, the non-zero ε_r with zero σ_{mean} was also reported in [7, 26] for mod. 9Cr–1Mo FM steel at 550 °C. It is possible to simulate such a material behavior with a constitutive model.

By modifying a constitutive model with such asymmetric properties (this is discussed in detail in Chapter 1), it is possible to simulate asymmetry of stress in symmetric strain-controlled LCF tests. Fig. 4.5a shows the simulated result: The peak tensile stress in the 100th cycle is 491.8 MPa while the peak compressive stress here is -509.3 MPa. The peak tensile and compressive strains are $\pm$0.5%. With the same model and the same parameter values, another simulation is performed with stress controlling, as shown in Fig. 4.5b and c. The stress amplitude is 500 MPa and σ_{mean} is zero. Owing to asymmetry, the stress-increasing half of the loop (blue curve in Fig. 4.5b and c) yields a larger change of strain than that in the stress-decreasing half of the loop (red curve in Fig. 4.5b and c). Consequently, the material has accumulated strain even with zero σ_{mean}. This is verified in the second stage of experiments at RT.

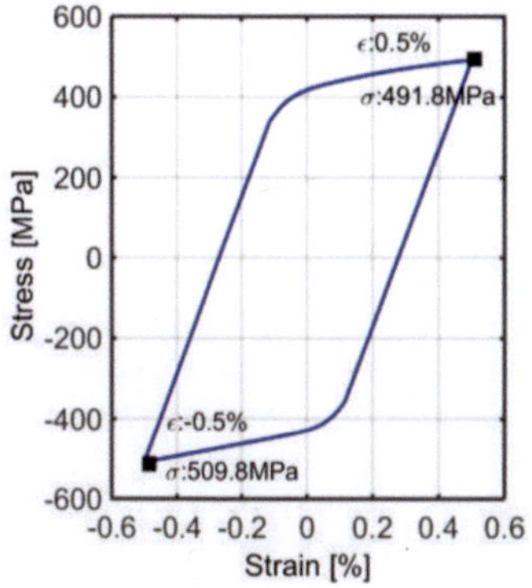

a)

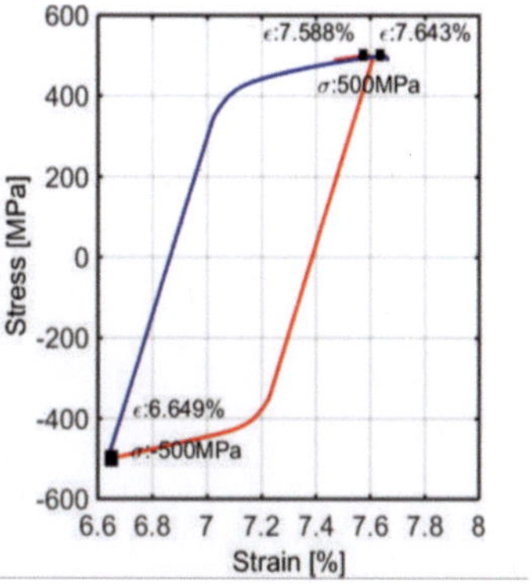

b)

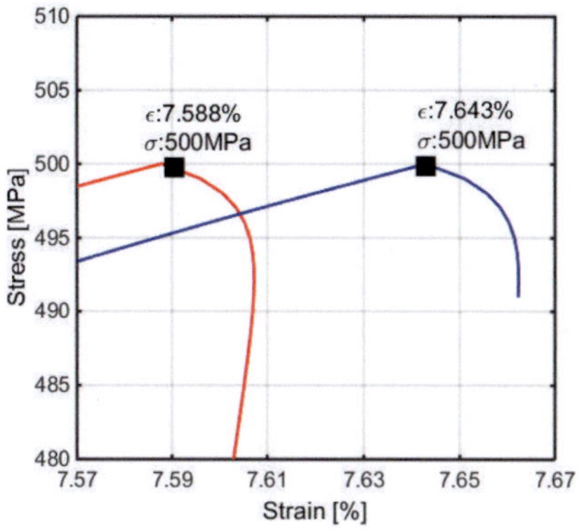

c)

Fig. 4.5: a) Simulation for a strain-controlled LCF test at RT performed with $\Delta\varepsilon$ =1.0%, the 100th cycle. b) Simulation for a stress-controlled ratcheting test with σ_{mean} = 0 MPa and σ_a = 500 MPa. c) Zoom of the peak tensile stresses of the sub-figure.

4.2 Results of Stress-controlled Ratcheting Tests

In the second stage of experiments at RT, uniaxial stress-controlled tests were carried out. Fig. 4.6 shows one of the tests performed with σ_{peak} = 500 MPa, σ_{mean} = 25 MPa, and $\dot{\sigma}$ = $\pm$50 MPa/s. The tests were performed with a triangle wave of stress. Hence, the magnitude of the stress rate was constant, as shown in Fig. 4.6a. The strain was yielded as a consequence. Fig. 4.6b illustrates the yielded strain in the first cycle.

Fig. 4.6c shows several hysteresis loops (1st, 250th, 500th, 750th, 1000th, and 1400th cycle) in this test. Although in the first cycle, the maximum strain reached only less than 0.5%, the strain reached more than 4.5% after the 1400th cycle. Therefore, ratcheting behavior exists at RT on P91.

Ratcheting is evaluated by plotting the mean strain (ε_{mean}) of each cycle together with its corresponding cycle number to show the rate of ratcheting. In other words, the speed of strain accumulation. Fig. 4.6d shows ε_{mean} versus number of cycles of this test.

Owing to cyclic softening, the strain range of hysteresis loop ($\Delta\varepsilon$) increases, as shown in Fig. 4.6e. The reduction of $\Delta\varepsilon$ in the initial cycles is owing to initial cyclic hardening, which corresponds to cyclic hardening in the initial 5–10 cycles of the strain-controlled LCF tests.

From an engineering point of view, the deformation of a component is not allowed to be larger than 5%. Concerning this reason, ratcheting tests can be ended after the strain reaches 5%. Since the measurement range of the extensometer is $\pm$10%, some ratcheting tests with fast $\acute{\varepsilon}_r$ were ended at strains larger than 5%. Owing to limited experimental time, some ratcheting tests with very low $\acute{\varepsilon}_r$ were ended at strains smaller than 5%.

In the following Sections 4.2.1–4.2.4, diagrams show results of various groups of ratcheting tests according to Table 3.3.

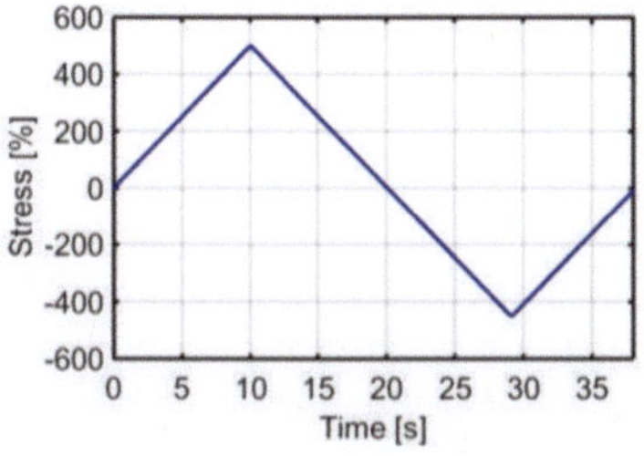

a) *1st cycle, stress v.s. time.*

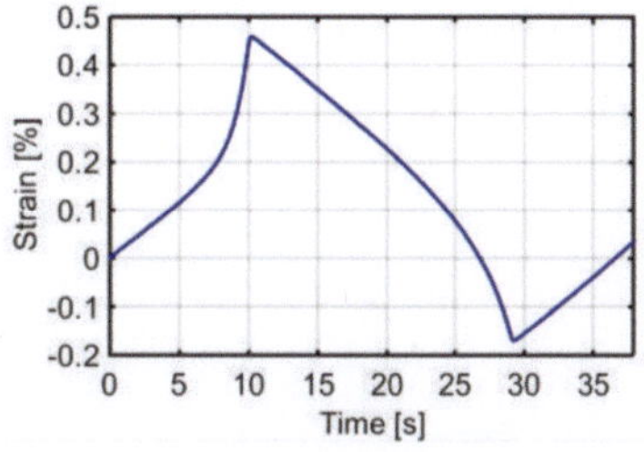

b) *1st cycle, strain v.s. time.*

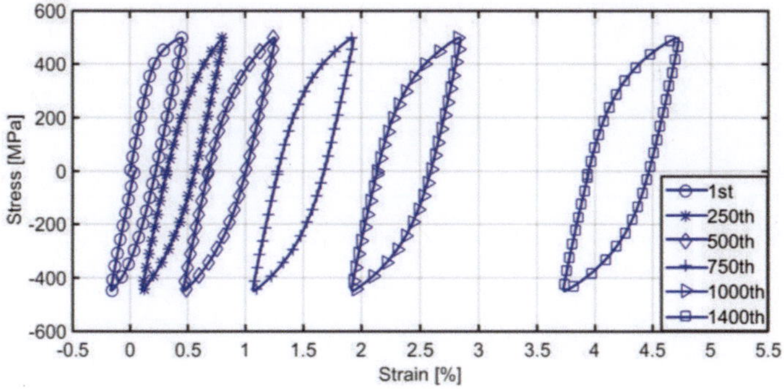

c) *Hysteresis loops of 1st,250th,500th,750th ,1000th,1400th cycle.*

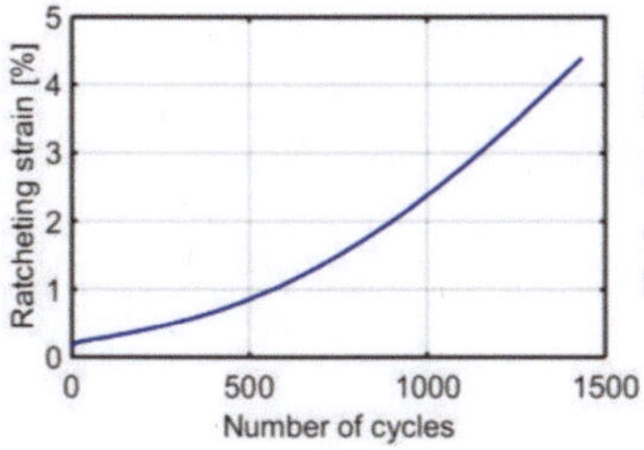

d) ε_r *vs. number of cycles*

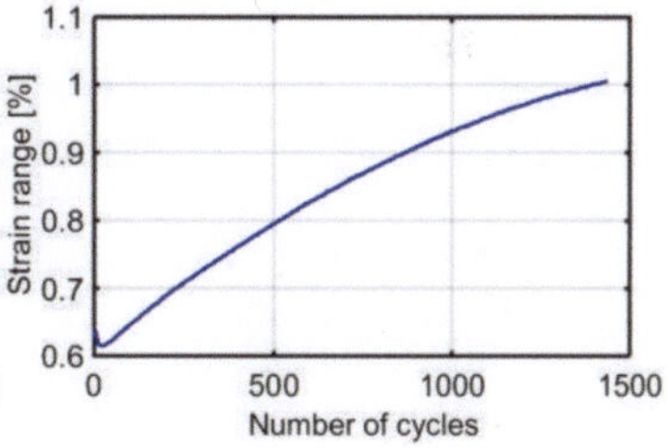

e) Strain ranges of hysteresis loops

Fig. 4.6: Ratcheting tests at RT performed with σ_{peak}=500 MPa, σ_{mean}=25 MPa, stress rate $\pm$50 MPa/s.

4.2.1 Influence of Peak Stress

According to Table 3.3a, eight experiments were carried out to explore the influence of peak stress on ratcheting. The σ_{mean} are all the same as 25 MPa and σ_{peak} range from 550 to 400 MPa. The ratcheting strains (mean strain in each cycle) are plotted together with the number of cycles in Fig. 4.7.

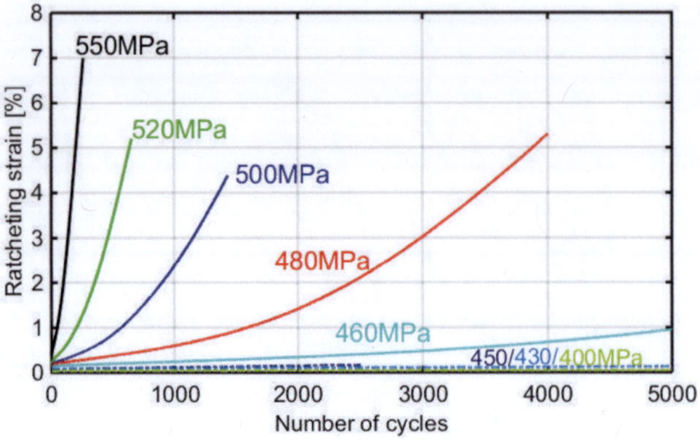

a)

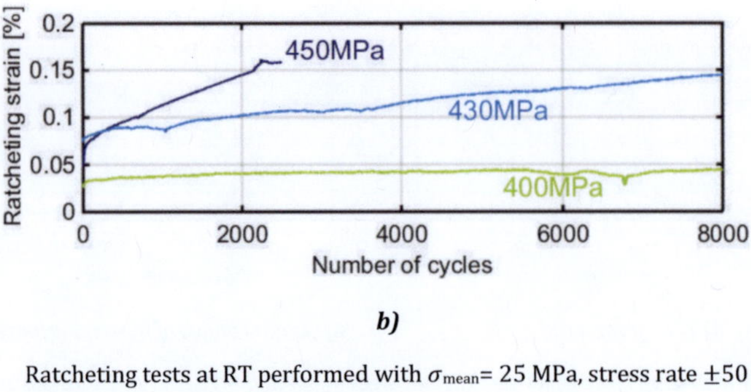

b)

Fig. 4.7: Ratcheting tests at RT performed with σ_{mean}= 25 MPa, stress rate ±50 MPa/s, various σ_{peak}, ε_r vs. number of cycles

In Fig. 4.7a, the ratcheting with $\sigma_{peak} \leq 450$ MPa is negligible compared to those tests performed with larger σ_{peak}. These results with σ_{peak} = 450, 430, and 400 MPa are plotted separately in Fig. 4.7b. With σ_{peak} = 550 MPa, the accumulated strain reaches 7% in the 238th cycle, while with σ_{peak} = 400 MPa, the maximum strain reaches only 0.2235% after 8000 loading cycles. Obviously the larger σ_{peak} leads to faster ratcheting, and the ratcheting behavior is negligible with σ_{peak} < 400 MPa.

The inelastic strain range $\Delta\varepsilon^{in}$ of each cycle is illustrated in Fig. 4.8. It is clear that $\Delta\varepsilon^{in}$ increases with increasing cycle number, which is due to cyclic softening. However, the cyclic softening in these stress-controlled tests does not show two stages as in the strain-controlled LCF tests (see Fig. 4.2c); instead, $\Delta\varepsilon^{in}$ decreases in the initial five cycles owing to initial hardening and then increases until the end of the test. Note that all eight tests shown in Fig. 4.7 and Fig. 4.8 ended without a macro crack.

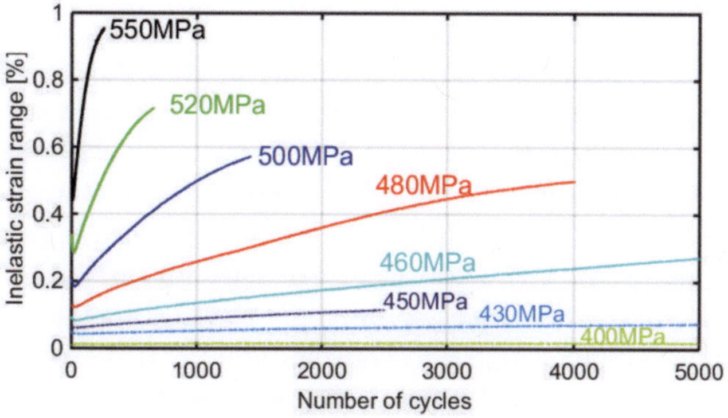

Fig. 4.8: Ratcheting tests at RT performed with σ_{mean}=25 MPa, stress rate ±50 MPa/s, various σ_{peak}, $\Delta\varepsilon^{in}$ vs. number of cycles

The average ratcheting rate $\acute{\varepsilon}_r$ are put together with their σ_{peak} in a log-linear plot (see Fig. 4.9) and a quasi-linear relationship can be seen. The $\acute{\varepsilon}_r$ shown in Fig. 4.9 are calculated as the average rates until the cycle in which the maximum strain reaches 4% or the cycle number achieves 2500, whichever comes first.

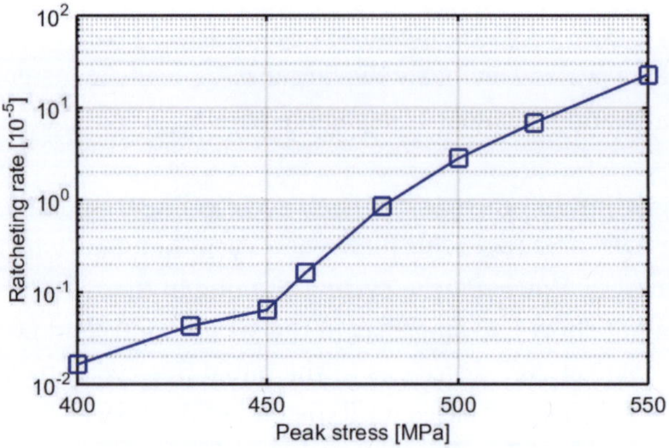

Fig. 4.9: Ratcheting tests at RT performed with σ_{mean} =25 MPa, stress rate ±50 MPa/s, average ratcheting rates vs. σ_{peak} in log-linear diagram

In the above-mentioned experiments, the focus is on the influence of peak tensile stress. Two additional verification tests were performed to check whether the influence of peak compressive stress is the same as that of peak tensile stress. One of these verification tests was performed with σ_{max} = 450 MPa and σ_{min} = -500 MPa, $\dot{\sigma}$ = ±50 MPa/s. Comparing to the test performed with a stress range of 500~-450 MPa, σ_a is the same, only σ_{mean} is changed from +25 MPa to -25 MPa. Another verification test was performed with a stress range of 500~-550 MPa, in comparison to the test performed with a stress range of 550~-500 MPa.

The results are illustrated in Fig. 4.10. The two loading cases with stress ranges of 550~-500 MPa and 500~-550 MPa have similar magnitudes of $\acute{\varepsilon}_r$; the loading cases with stress ranges of 500~-450 MPa and 450~-500 MPa also have similar magnitudes of $\acute{\varepsilon}_r$. However, the case with 450~-500 MPa has a lower $\acute{\varepsilon}_r$ (absolute value) than that in the loading case of 500~-450 MPa.

A similar difference can be found between the two tests performed with stress ranges of 550~-500 MPa and 500~-550 MPa. These small differences between the absolute values of $\acute{\varepsilon}_r$ show that more accumulated strain is induced with tensile σ_{mean} than with the same magnitude of compressive σ_{mean}. Together with the asymmetry of the σ_{peak} in the symmetric strain-controlled LCF test reported in Section 4.1, they support the assumption of asymmetry of material strength under tension and compression.

One confusion can be eliminated after these two verification tests shown in Fig. 4.10: The so-called "influence of peak stress", as the title of the current section, is specifically the "influence of maximum absolute value of stress". In other words, if peak compressive stress has a larger absolute value than that of peak tensile stress, then the $\acute{\varepsilon}_r$ is mainly decided by the peak compressive stress. However, since most ratcheting tests performed in the current work have larger absolute values of peak tensile stresses than the absolute values of peak compressive stresses, the term "peak stress" or "maximum stress" means "peak tensile stress" ($\sigma_{peak} = \sigma_{max}$) if given without a specific explanation.

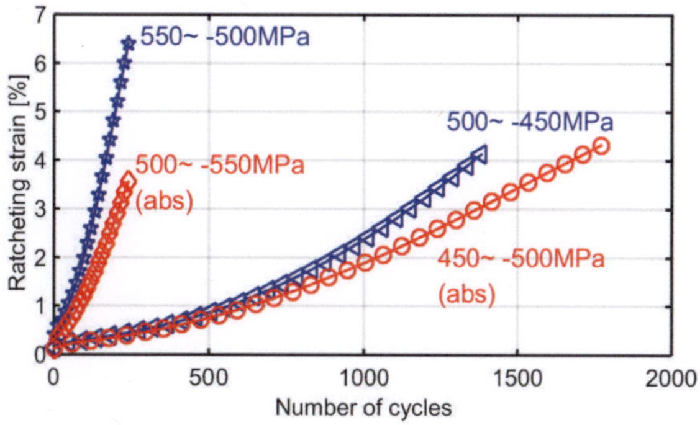

Fig. 4.10: Two groups of ratcheting tests at RT performed with symmetric σ_{mean}, ε_r vs. number of cycles.

4.2.2 Influence of mean stress/stress ratio.

In stress-controlled tests, if the σ_{peak} are the same, one stress ratio (R) corresponds to one mean stress σ_{mean}

$$\sigma_{mean} = \frac{\sigma_{peak} + \sigma_{min}}{2} \tag{4-1}$$

$$R = \frac{\sigma_{min}}{\sigma_{peak}} \tag{4-2}$$

With (4-1) and (4-2), we have

$$R = \frac{2\sigma_{mean} - \sigma_{peak}}{\sigma_{peak}} \tag{4-3}$$

Hence "the influence of mean stress" and "the influence of stress ratio" are identical with the same σ_{peak}.

The experiments were performed according to Table 3.3 b). ε_r versus number of cycles is plotted in Fig. 4.11. They have the same peak tensile stress of 500 MPa and the same stress rate of $\pm$50 MPa/s. The tests with σ_{mean} = 100 and 250 MPa are plotted separately in Fig. 4.11 b) for their relatively low $\acute{\varepsilon}_r$.

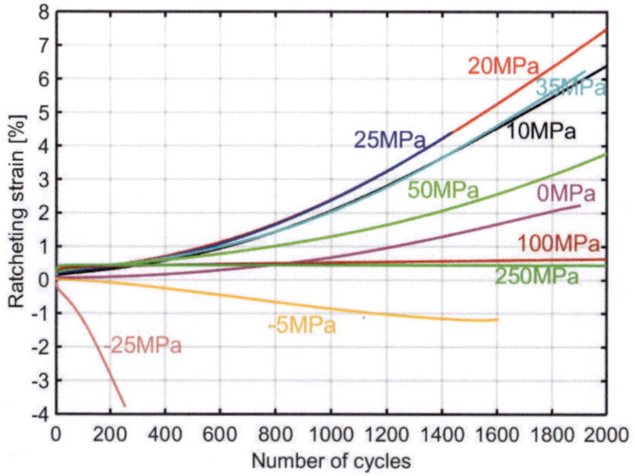

a)

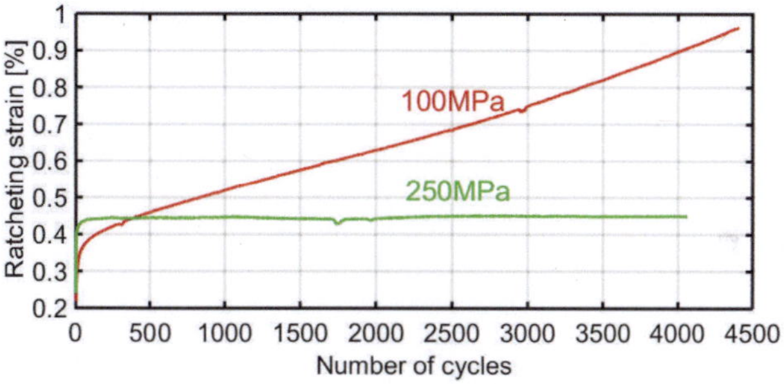

b)

Fig. 4.11: Ratcheting tests at RT performed with σ_{peak}=500 MPa, stress rate $\pm$50 MPa/s, various σ_{mean}, ε_r vs. number of cycles.

$\Delta\varepsilon^{in}$ versus number of cycles is plotted in Fig. 4.12. The evolution of $\Delta\varepsilon^{in}$ is similar to that shown in Fig. 4.8, with an initial decrease followed by an even and smooth increase until the end of the test. As shown in Fig. 4.12, a larger stress range leads to a higher increase in the rate of $\Delta\varepsilon^{in}$. For instance, the largest stress range is with σ_{mean} = -25 MPa (σ_{max} = 500 MPa, σ_{min} = -550 MPa), which has the highest increase in rate of $\Delta\varepsilon^{in}$. No macro crack was observed on the specimens for these 10 ratcheting tests shown in Fig. 4.11 and Fig. 4.12.

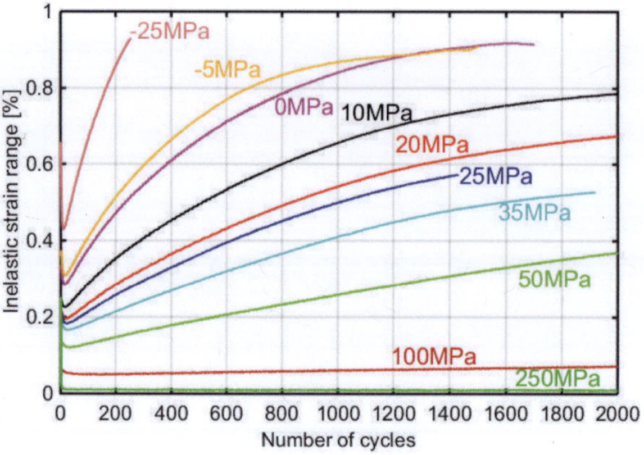

Fig. 4.12: Ratcheting tests at RT performed with σ_{peak} = 500 MPa, stress rate ±50 MPa/s, various σ_{mean}, $\Delta\varepsilon^{in}$ vs. number of cycles.

Fig. 4.13 shows the relation between average ε_r' and stress ratio. The average ε_r' before the 1250th cycle of the experiments together with their corresponding stress ratios are plotted in this diagram. The case with stress range 500--550 MPa is not shown in Fig. 4.13 because it did not reach 1250th cycle before buckling.

Unlike the influence of peak stress, the ratcheting is not accelerated by an increase of mean stress. As shown in Fig. 4.11a, ε_r' reaches its maximum with σ_{mean} = 20 MPa. The largest mean stress of 250 MPa, on the contrary, leads to the minimum magnitude of ε_r'. As shown in Fig. 4.13, ε_r' reaches the maximum with the stress ratio at around -0.9.

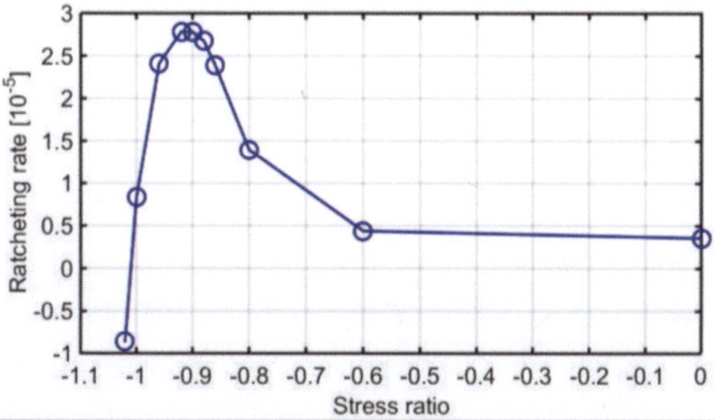

Fig. 4.13: Ratcheting tests at RT performed with σ_{peak}=500MPa, stress rate ±50MPa/s, various σ_{mean}, average ratcheting rates (until 1250th cycle) vs. stress ratios.

This phenomenon can be explained by Fig. 4.14: The hysteresis loops include mostly elastic regions if σ_{mean} = 250 MPa, which means the inelastic strain is too small to induce ratcheting. Consequently, ratcheting with the same σ_{peak} and larger σ_{mean} has less inelastic strain, which leads to a lower $\acute{\varepsilon}_r$. An intuitive explanation can be stated as follows: $\acute{\varepsilon}_r$ is controlled by the difference between the change of inelastic strain ($|\Delta\varepsilon^{in}|$) during the stress-increasing and stress-decreasing half in each cycle. With stress ratio $-1 < R < -0.9$, both $|\Delta\varepsilon^{in}|$ during the stress-increasing and -decreasing half are large and $|\Delta\varepsilon^{in}|$ during stress-increasing half is even larger than that during the stress-decreasing half. At $R \cong -0.9$, this difference between the two $|\Delta\varepsilon^{in}|$ reaches the maximum. When $R > -0.9$ and approaches zero, both $|\Delta\varepsilon^{in}|$ during stress-increasing and -decreasing half decrease and the difference decreases correspondingly until zero, which leads to a decrease of $\acute{\varepsilon}_r$.

However, $\acute{\varepsilon}_r$ in the test performed with σ_{peak} = 500 MPa and σ_{mean} = 250 MPa is still not exactly zero since the material response is not completely elastic. As shown in Fig. 4.14a, the difference between the loops of the 1st and 500th cycles is obvious, but the loops of the 500th and 4000th cycles are almost the same and are both composed only of an elastic region.

Fig. 4.14b shows the inelastic strain ranges $\Delta\varepsilon^{in}$ in each cycle and Fig. 4.14c illustrates those in the initial 30 cycles. It is clear that $\Delta\varepsilon^{in}$ decreased rapidly in the initial cycles and stayed constant afterwards, with a value of only 0.01%. Therefore, the elastic shake down happens in this case.

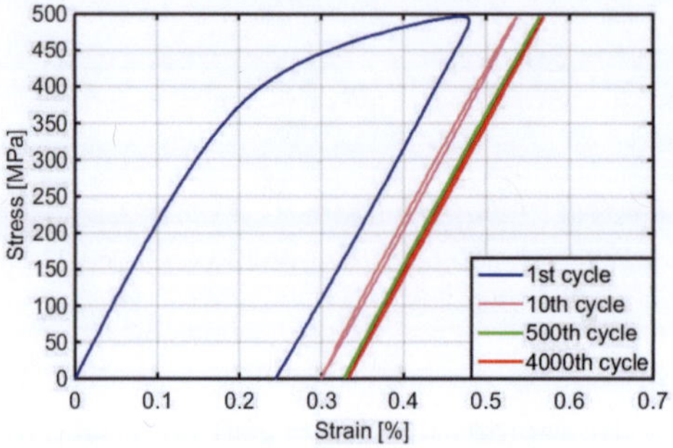

***a)** Hysteresis loops of the 1st, 500th and 4000th cycle*

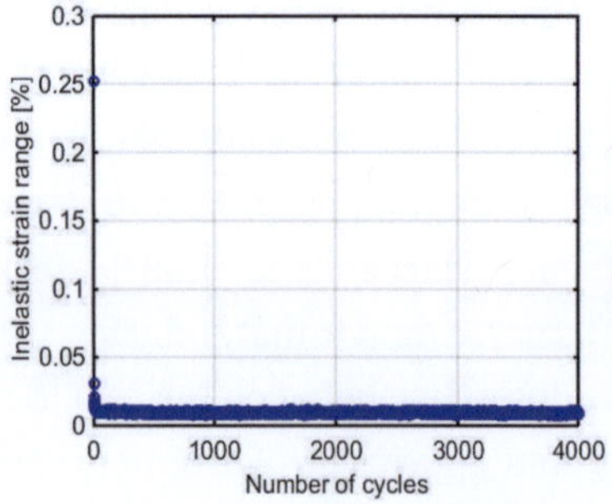

***b)** $\Delta\varepsilon^{in}$ until 4000th cycle*

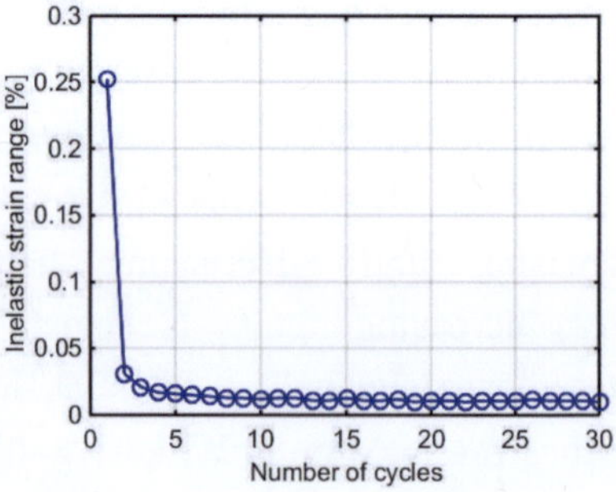

***c)** $\Delta\varepsilon^{in}$ in the initial 30 cycles*

Fig. 4.14: Ratcheting test at RT performed with σ_{peak} = 500 MPa, σ_{mean} = 250 MPa, stress rate ±50 MPa/s.

4.2.3 Ratcheting with zero mean stress

In symmetric strain-controlled LCF tests, asymmetry of peak tensile and compressive stress is observed. This can yield positive ratcheting, as discussed in Section 4.1. Based on this assumption, a group of tests was carried out according to Table 3.3c, the results of which are shown in Fig. 4.15. In the test performed with σ_a = 450 MPa, $\acute{\varepsilon}_r$ is relatively negligible, hence it is separately shown in Fig. 4.15b.

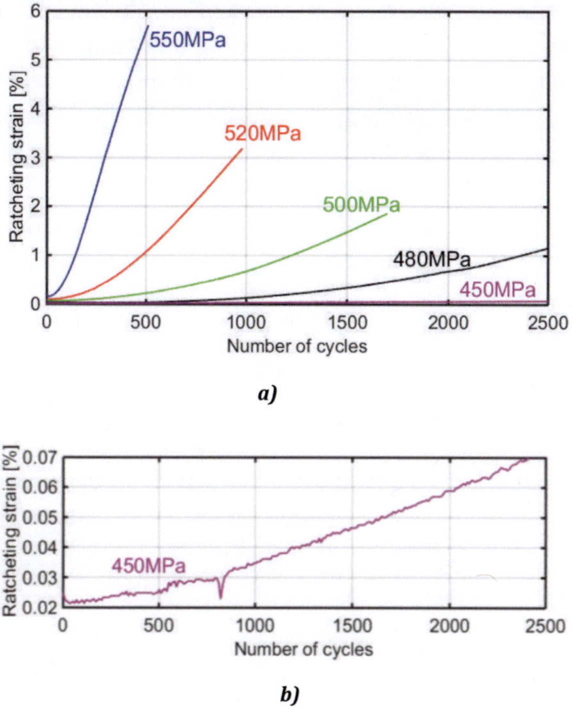

Fig. 4.15: Ratcheting tests at RT performed with zero σ_{mean}, stress rate ±50 MPa/s, various σ_a, ε_r vs. number of cycles.

$\Delta\varepsilon^{in}$ versus number of cycles is plotted in Fig. 4.16. Only the tests performed with σ_a = 520 MPa were ended with macro cracks, and the corresponding $\Delta\varepsilon^{in}$ increases in the last stage of this test. The other four tests were ended without macro cracks. $\Delta\varepsilon^{in}$ in the test performed with σ_a = 550 MPa showed an obvious decrease of the rate and $\Delta\varepsilon^{in}$ even decreased in the last stage of the test. This indicates that the cyclic softening of the material under stress-controlled loading can be progressively saturated and followed by further hardening.

Another possible reason for the decrease of $\Delta\varepsilon^{in}$ in the last stage of the test performed with σ_a = 550 MPa is that, at very large deformation (ε > 5%), the effect of asymmetry of tensile and compressive strength is less. However, this phenomenon only appeared once and was not observed in other RT ratcheting tests.

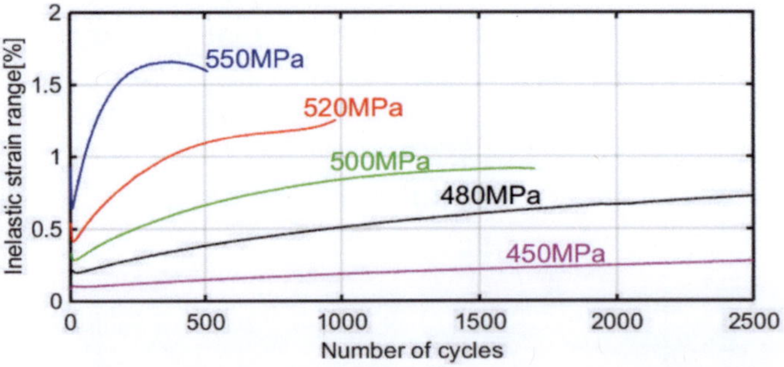

Fig. 4.16: Ratcheting tests at RT performed with zero σ_{mean}, stress rate ±50 MPa/s, various σ_a, and $\Delta\varepsilon^{in}$ vs. number of cycles.

It is clear that $\dot{\varepsilon}_r$ under symmetric stress-controlled loading is higher with larger peak tensile stress, which is also stress amplitude σ_a since σ_{mean} is zero. As shown in Fig. 4.17, in the log-linear plot of $\dot{\varepsilon}_r$ versus σ_a, a quasi-linear line can be found. The $\dot{\varepsilon}_r$ shown in Fig. 4.17 are calculated as the average rates until the cycle in which the maximum strain reaches 2% or the cycle number achieves 1500, which ever comes first.

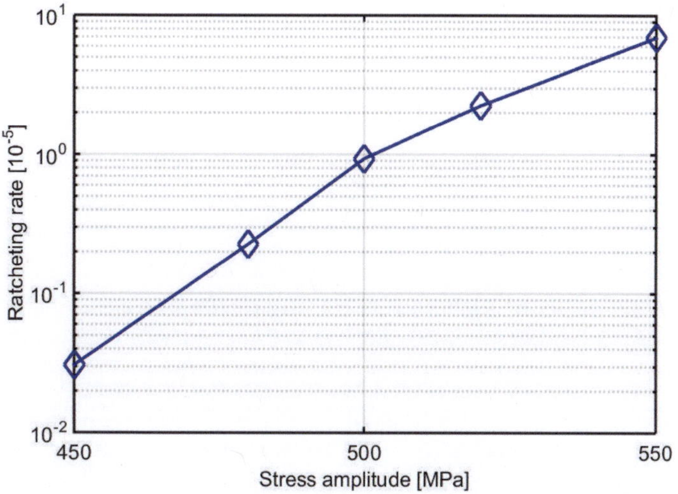

Fig. 4.17: Ratcheting tests at RT performed with zero σ_{mean}, stress rate ± 50 MPa/s, various σ_a, average ratcheting rates vs. σ_a in log-linear diagram.

There are now three phenomena for P91 at RT:

1. The asymmetry of peak tensile and compressive stress in strain-controlled LCF tests (see Fig. 4.4).
2. Lower magnitude of $\dot{\varepsilon}_r$ with compressive σ_{mean} (see Fig. 4.10).
3. Positive $\dot{\varepsilon}_r$ with zero σ_{mean} (see Fig. 4.17).

These three phenomena together lead to the conclusion that at room temperature the strength of P91 is higher under compression than under tension.

4.2.4 Influence of stress rate

Owing to the possible visco-plasticity of P91 steel at RT, the influence of time on ratcheting behavior was tested in experiments, including tests performed with different $\dot{\sigma}$ (stated in this section) and tests performed with different hold times (stated in Section 4.2.4). Two tests with $\dot{\sigma}$ = $\pm$10 and $\pm$250 MPa/s were performed. Together with the test performed with $\dot{\sigma}$ = $\pm$50 MPa/s in the previous group, these results are shown in Fig. 4.18. The σ_{peak} and σ_{mean} are the same in these three tests, as listed in Table 3.3d.

It is clear that visco-plasticity plays a role at RT, as shown in Fig. 4.18, since lower $\dot{\sigma}$ leads to higher $\acute{\varepsilon}_r$. According to the average $\acute{\varepsilon}_r$ before the cycle in which the maximum strain reaches 4%, the test performed with $\pm$10 MPa/s has an 18.3% higher $\acute{\varepsilon}_r$ than the test performed with $\pm$50 MPa/s, and the test performed with $\pm$250 MPa/s has a 22.3% lower $\acute{\varepsilon}_r$ than the test performed with $\pm$50 MPa/s. Hence, a five times higher $\dot{\sigma}$ leads to an approximately 20% lower $\acute{\varepsilon}_r$.

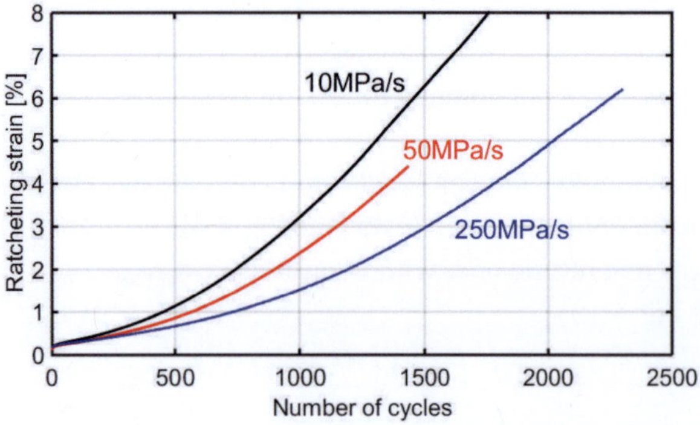

Fig. 4.18: Ratcheting tests at RT performed with σ_{peak} = 500 MPa, σ_{mean} = 25 MPa, various stress rates, ε_r vs. number of cycles.

The inelastic strain range of each cycle is plotted with corresponding cycle number in Fig. 4.19. It is clear that the higher stress rate leads to lower $\Delta\varepsilon^{in}$ owing to visco-plasticity. The evolution of $\Delta\varepsilon^{in}$ with each stress rate is similar to that in the other ratcheting test at RT (see

Fig. 4.8, Fig. 4.12, and Fig. 4.16): After a short initial decrease, $\Delta\varepsilon^{in}$ increases evenly and smoothly but with a decreasing increase rate until the end of the test.

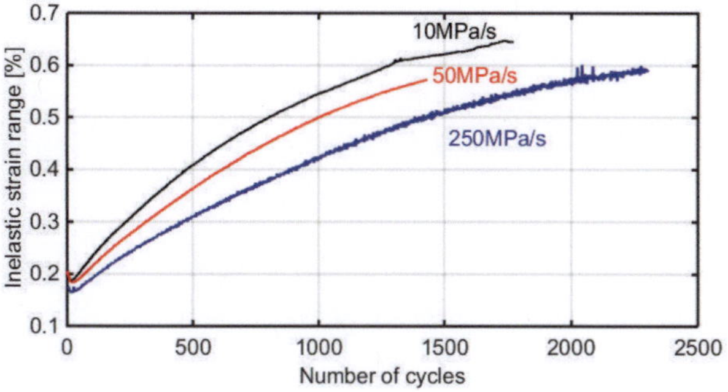

Fig. 4.19: Ratcheting tests at RT performed with σ_{peak} = 500 MPa, σ_{mean} = 25 MPa, various stress rates, $\Delta\varepsilon^{in}$ vs. number of cycles.

4.2.5 Influence of hold time

As planned in Table 3.3e, two tests with 10 min hold times at peak tensile and compressive stresses were performed to discover the role of time/visco-plasticity at RT. Together with the test performed with no hold time in the previous group, the results are shown in Fig. 4.20. The σ_{peak}, σ_{mean} and $\dot{\sigma}$ were the same in these three tests.

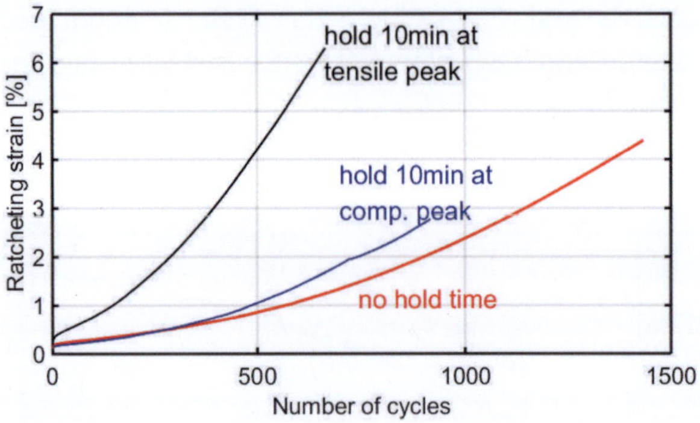

Fig. 4.20: Ratcheting tests at RT performed with σ_{peak} = 500 MPa, σ_{mean} = 25 MPa, stress rate ±50 MPa/s, various hold time types, ε_r vs. number of cycles.

The 10min hold times at tensile peaks (500 MPa) dramatically accelerated the ratcheting while hold times at compressive peaks (-450 MPa) scarely accelerated the ratcheting.

After analysis of the two tests with hold times at the tensile and compressive peaks, the inelastic strains during the hold time in each cycle can be found, as illustrated in Fig. 4.21. As can be seen, the inelastic strains at tensile and compressive peaks agree with each other. The inelastic strain during the first cycle of the test with 10 min hold time at tensile peak is around 0.115%. The inelastic strains in the initial approximately 30 cycles decrease rapidly and then slightly increase again until the end of the test owing to cyclic softening. Because of the limit of accuracy, only conservative calculations of the accumulated inelastic strains can be acquired, which are 16.70% and -15.63% at tensile and compressive hold times, respectively. Concerning the total accumulated inelastic strain, the accumulated partial inelastic strain formed during the stress-changing process was -12.49% in the test performed with 10 min hold times at tensile peaks, while this accumulated strain of the test performed with 10 min hold times at compressive peaks was 16.68%.

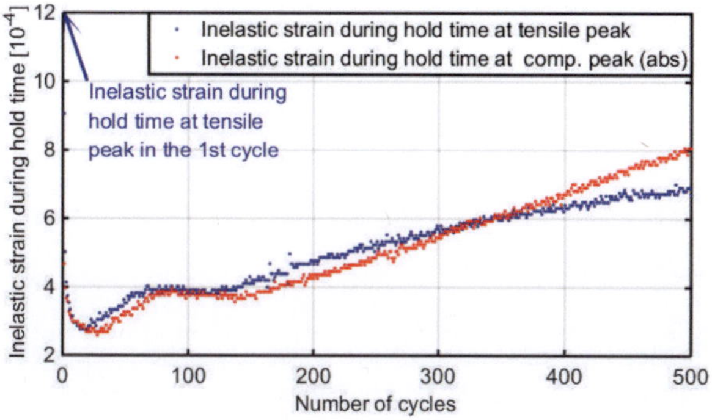

Fig. 4.21: Ratcheting tests at RT performed with σ_{peak} = 500 MPa, σ_{mean} = 25 MPa, stress rate ±50 MPa/s, inelastic strain during hold times at tensile & compressive peaks vs. number of cycles.

An additional test was performed with the same equipment as the ratcheting tests, with stress of 500 MPa. The result is illustrated in Fig. 4.22. In spite of the initial stage of fast increase of inelastic strain, the strain increase after the first hour until the 24th hour is negligible. Therefore, the accumulated inelastic strain during hold times in the ratcheting test is much higher than that in the test with static stress of 500 MPa, because the mechanical behavior during hold times is similar to the initial stage in the test with static stress. The inelastic strain in the first 10 min in the test with static stress is 0.18%, which is comparable with 0.12% during the 10 min hold time at tensile peak in the first cycle of the ratcheting test.

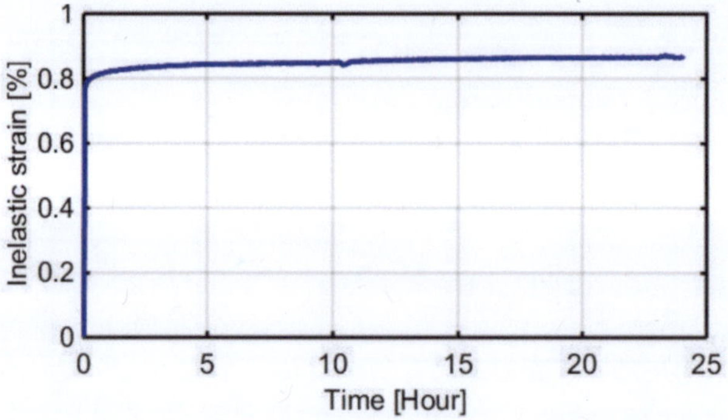

Fig. 4.22: Test at RT performed with σ=500MPa, inelastic strain vs. time.

To understand why the accumulated inelastic strain during the stress-changing process is negative (-12.49%) in the test performed with hold times at tensile peaks, the hysteresis loops of three tests are plotted in Fig. 4.23. The three loops are aligned by setting the center as zero. As can be seen, in spite of a slight total inelastic strain increment (ratcheting), the material response to the cyclic stress-controlled loading has a tendency to close the hysteresis loop. This tendency is much clearer in the case with hold time at the compressive peak: As shown in Fig. 4.23, the change of inelastic strain during the hold time is -0.080%, while the total inelastic strain increment during this cycle (ratcheting) is only 0.003%. This means the partial inelastic strain formed during the stress-changing process compensates most of the partial inelastic strain increment during the hold time. Owing to this tendency, $\acute{\varepsilon}_r$ of the test performed with hold time at tensile peak is only around five times higher than that without hold time, and the test performed with hold times at compressive peaks still has positive ratcheting.

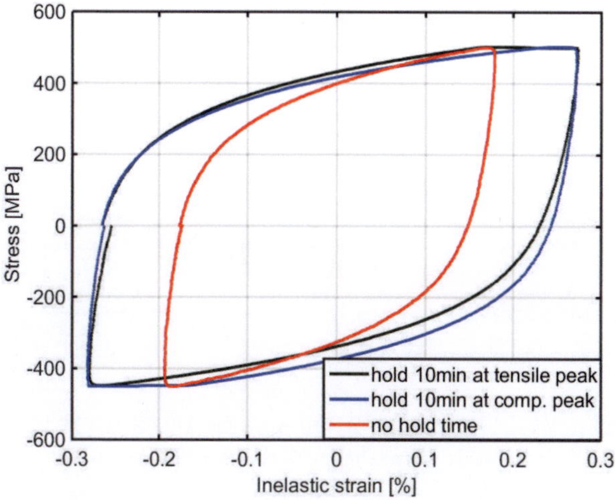

Fig. 4.23: Ratcheting tests at RT performed with σ_{peak} = 500 MPa, σ_{mean} = 25 MPa, stress rate ±50 MPa/s, stress vs. inelastic strain (normalized) at 500th cycle.

This tendency to close the loop can also be approved by the decrease of $\acute{\varepsilon}_r$ in the initial cycles in the ratcheting tests, when cyclic softening has not played a key role: $\acute{\varepsilon}_r$ in each cycle versus cycle number is plotted in Fig. 4.24. Those $\acute{\varepsilon}_r$ in the initial cycles are marked with larger markers. As can be seen, $\acute{\varepsilon}_r$ decreases rapidly in the initial cycles, which means the hysteresis loop tends to close in the initial stage of the ratcheting tests. $\acute{\varepsilon}_r$ increases slightly in the following cycles until the end of the test, owing to cyclic softening. This phenomenon also exists in the other ratcheting tests.

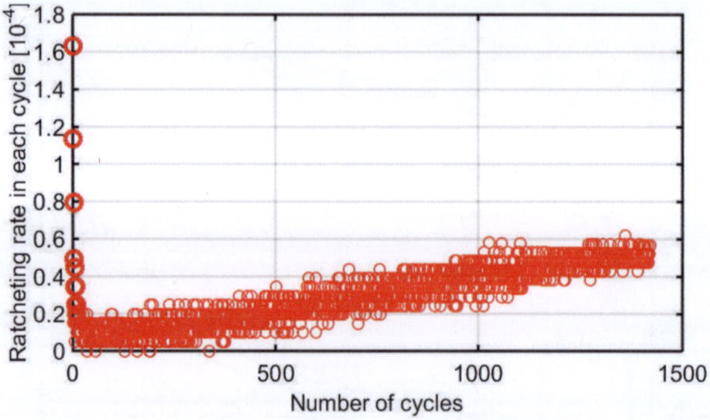

Fig. 4.24: Ratcheting test at RT performed with σ_{peak} = 500 MPa, σ_{mean} = 25 MPa, stress rate ±50 MPa/s, no hold time, ratcheting rate vs. number of cycles.

4.2.6 Diameter Check

After the experiments at room temperature, the diameters of the specimens in different axial positions were checked. Fig. 4.25 shows some examples of the tested specimens. The three marked diameters on each specimen include two near the end of gage length and one in the middle.

As shown in Fig. 4.25, the three measured diameters on each specimen are similar, hence no necking occurs. Further, in the test of stress range 550~-500 MPa, the average of the three measured diameters was 8.50 mm while the initial diameter of the received specimen was 8.80 mm. The shrinkage of the cross section is 1/1.072, which approximately corresponds to the 7.4% of strain. In the other tests, the shrinkage of the specimen also corresponds to each final strain. Hence the prerequisite to apply the so called "true-stress-controlled ratcheting test" is verified, which is that the shrinkage of cross section corresponds to the real time strain, based on the assumption of no volumetric plastic strain during tests.

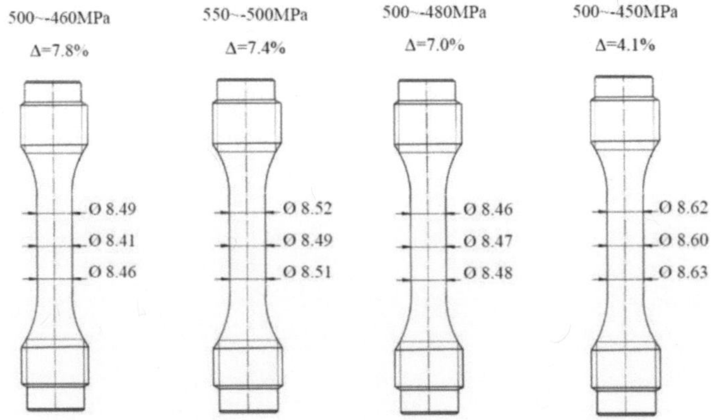

Fig. 4.25: Diameter check of several specimens in RT ratcheting tests.

5 Experiments at 550° C

The previous chapter reported the material responses under symmetric uniaxial strain-controlled cyclic loading and under asymmetric/symmetric uniaxial stress-controlled cyclic loading at room temperature.

Various unique phenomena were found on P91 at room temperature, such as cyclic softening and asymmetry of material strength under tension and compression. It is interesting to investigate the material responses at high temperature (550 °C) to check whether these phenomena still exist.

In this chapter, the results of experiments at 550 °C are presented. The hydraulic machine, measure and control equipment, and software are the same as at those used at RT, as presented in Section 3.1. The difference is an additional oven to heat the specimens together with the specimen-holder for the oven. As reviewed in Chapter 1, oxidation effect on cyclic softening and ratcheting can be ignored [82-84], and crack initiation and propagation is not taken into account.

The planned experiments are listed in Section 3.3, including strain-controlled LCF tests in Section 3.3.1 and stress-controlled tests in Section 3.3.1. The results of the strain-controlled tests are presented in Section 5.1, and Section 5.2 reports the results of stress-controlled tests, including the influences of various factors on ratcheting, such as σ_{peak}, σ_{mean}, $\dot{\sigma}$, and hold time. The ratcheting results with zero σ_{mean} at 550 °C are shown in Section 5.2.3, in comparison to those at RT reported in Section 4.2.3.

5.1 Results of Strain-controlled Tests

According to Table 3.4, five symmetric strain-controlled LCF tests were performed. The results are essentially the same as those shown in Fig. 4.1a–c.

The main difference between results at RT and 550 °C is obviously lower induced stresses at 550 °C. Fig. 5.1a shows the hysteresis loops of the 1st, 200th, and 400th cycles of the LCF test performed with $\Delta\varepsilon$ = 1.5%. Compared with Fig. 4.1d, it was found that both the initial σ_{peak} and those during saturation stage were much lower at 550 °C than those at RT, since the material strength is much lower at 550 °C than at RT.

Another difference is that the initial cyclic hardening is negligible at 550 °C; the peak tensile stress monotonically decreases either after the 1st cycle ($\Delta\varepsilon$ = 1.5 and 1.2%) or after the 2nd cycle ($\Delta\varepsilon$ = 1.0 and 0.8 and 0.6%). Fig. 5.1b shows the peak tensile stresses in the initial 15 cycles in the LCF test performed with $\Delta\varepsilon$ = 1.5%, compared with Fig. 4.1e for the case at RT.

One more difference is the obviously shorter lifetime. Comparing Fig. 5.1c and Fig. 4.1f, the cycle number until macro cracks appear at 550 °C is around 400, compared with around 1100 cycles at RT.

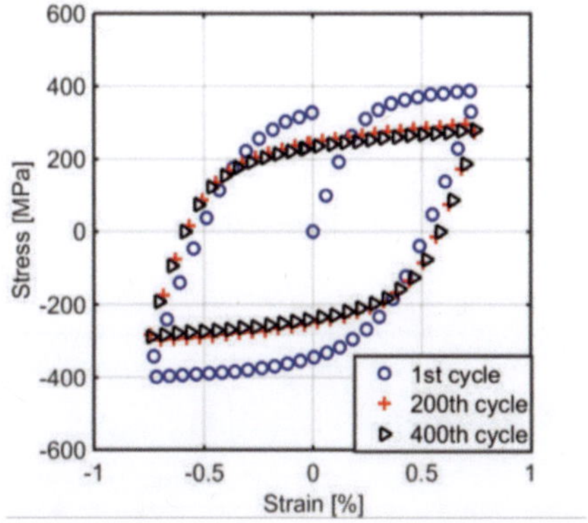

***a)** Hysteresis loops of 1st, 200th and 400th cycle.*

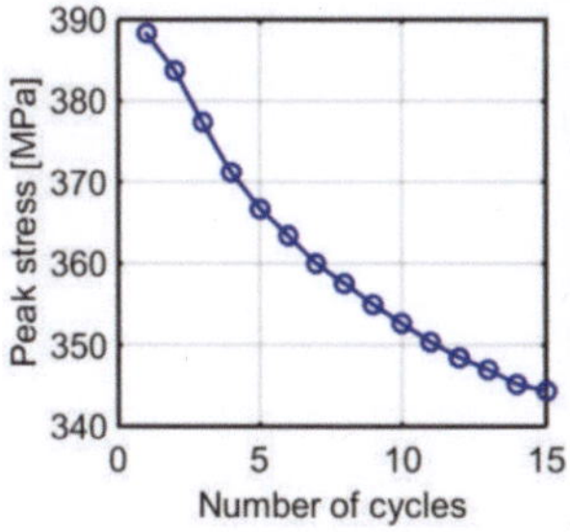

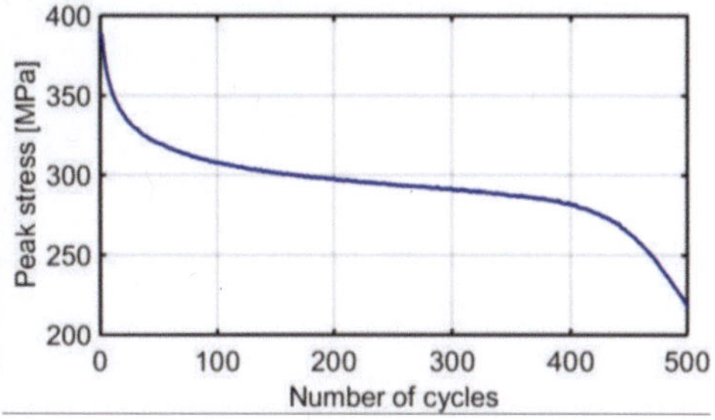

***b)** σ_{peak} in initial 15 cycles.* ***c)** σ_{peak} in cycles till fracture*

Fig. 5.1: Symmetric strain-controlled LCF tests at 550 °C, $\Delta\varepsilon$=1.5%.

The lifetime in LCF test at 550 °C is clearly defined, which is different from that for RT cases, because the peak tensile stresses in all five LCF tests decrease smoothly cycle by cycle after the macro cracks appear, since the gage length of the extensometer is 20 mm and the macro crack always appears within this gage length. As illustrated in Fig. 5.2, the cycle number of lifetime, or cycle number to failure, is defined as the cycle during which σ_{peak} has decreased by 10% from that predicted by extrapolation of the saturation curve, as in Saad et al. [101].

Hysteresis loops of initial cycles and cycles at the half lifetime of the five LCF tests are illustrated in Fig. 5.3a and b, respectively, and the peak tensile stresses are illustrated in Fig. 5.3c. As same as at RT, the peak tensile stress decreases rapidly in the initial 1/4 lifetime and the saturation stage follows afterwards until macro cracks appear.

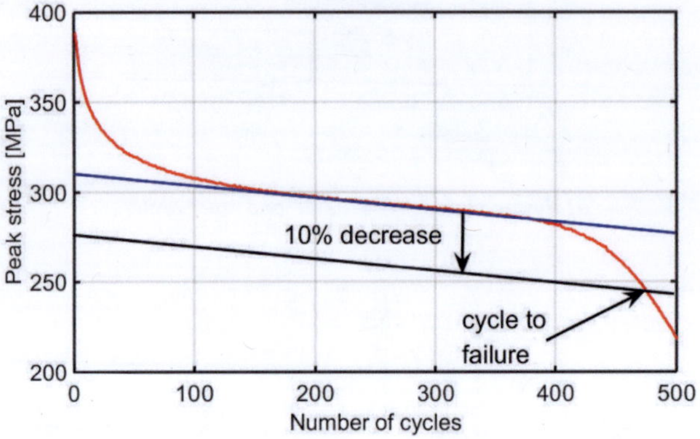

Fig. 5.2: Definition of lifetime in LCF tests at 550 °C. Strain-controlled LCF test with strain range 1.5%.

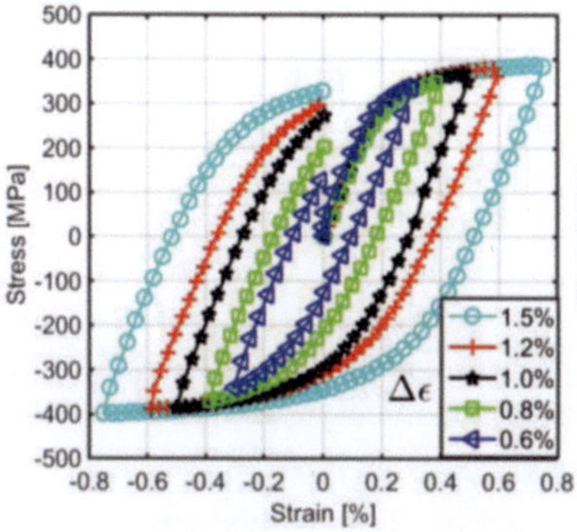

a) Hysteresis loops of the first cycle

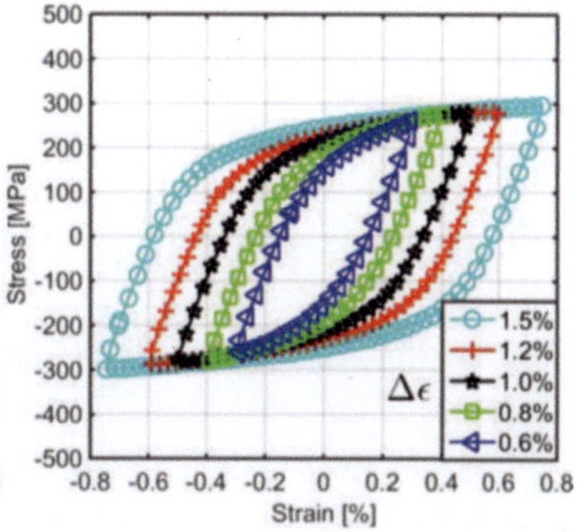

b) Hysteresis loops of the cycle at $N_f/2$

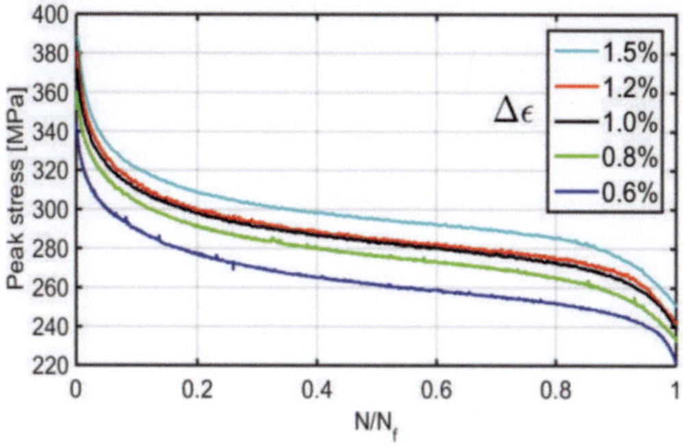

c) Peak tensile stresses vs. normalized number of cycles.

Fig. 5.3: Strain-controlled LCF tests at 550 °C performed with strain ranges 0.6% ~ 1.5%.

The grades of cyclic softening at two temperatures are also compared. By calculating the ratio between the engineering peak tensile stress in the middle of the saturation stage and the maximum engineering peak tensile stress during whole lifetime, it was found that the grade of cyclic softening is larger at 550 °C: The ratio is 0.88 ± 0.01 at RT and 0.76 ± 0.01 at 550 °C.

The magnitudes of peak compressive stresses also decrease as peak tensile stresses. There were also differences of 2~5 MPa between peak tensile and compressive stresses in all cycles. As shown in Fig. 5.4, the red curves indicate the absolute value of peak compressive stresses and the blue curves indicate peak tensile stresses.

By converting the σ_{eng} in Fig. 5.4 into σ_{true} in Fig. 5.5, pure material responses are clear: The absolute values of true peak compressive stresses are not larger than the true peak tensile stresses, both in the initial rapid decreasing stages of peak stresses and in the saturation stage. The differences appear only after the appearance of macro cracks.

It is clear that the asymmetry of σ_{peak} in the symmetric strain-controlled LCF at 550 °C is negligible, comparing to the asymmetry at RT. It is assumed that the material strength under tension and compression is the same at 550 °C. Therefore, it is interesting to predict what can happen in the following stress-controlled tests: is there still positive ratcheting under symmetric loading as at RT?

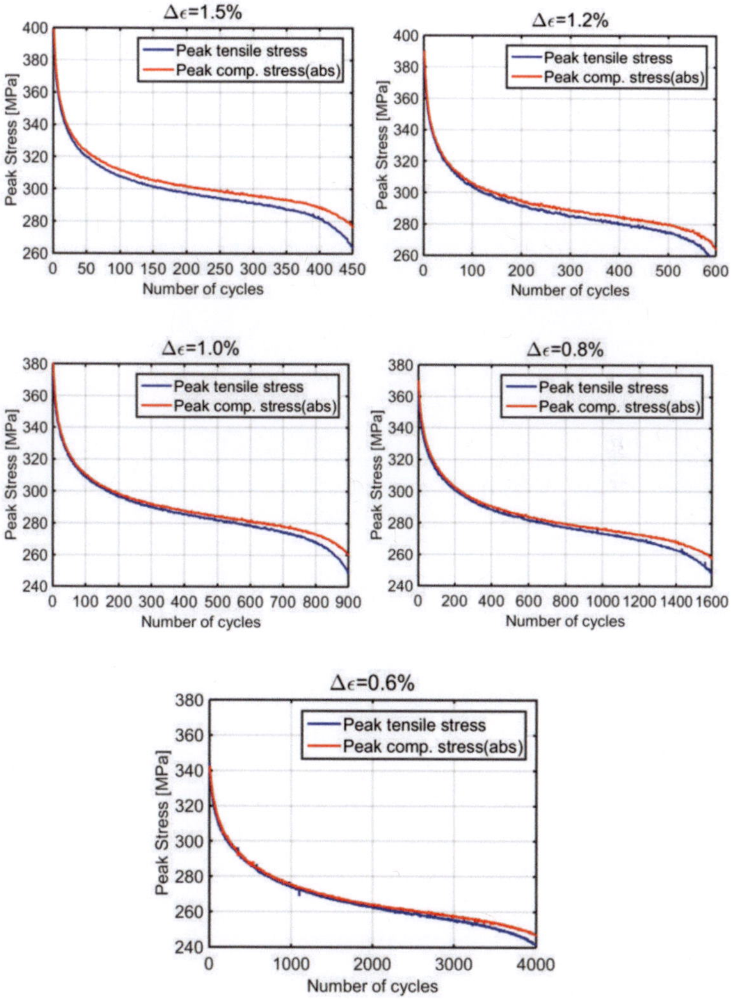

Fig. 5.4: Engineering peak compressive stresses (abs.) compared with engineering peak tensile stresses at 550 °C.

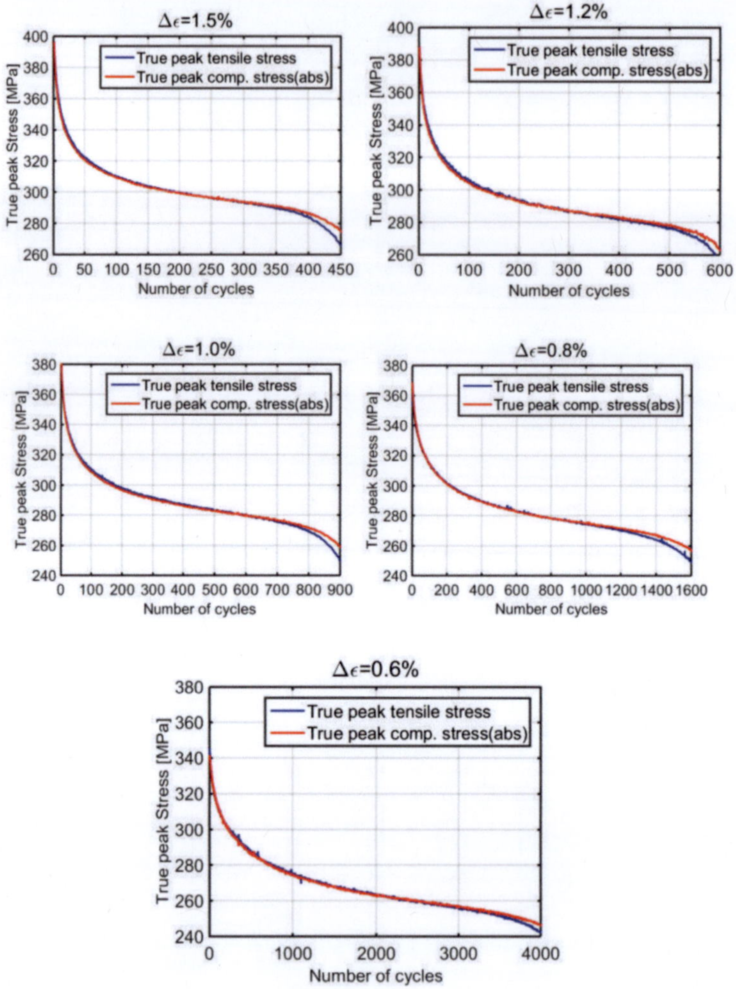

Fig. 5.5: True peak compressive stresses (abs.) compared with true peak tensile stresses at 550 °C.

In addition, due to the non-negligible stress relaxation in strain-controlled tests at high temperature, another two LCF tests were performed with hold time at peak strain under tension and under compression. The results are presented in Fig. 5.6.

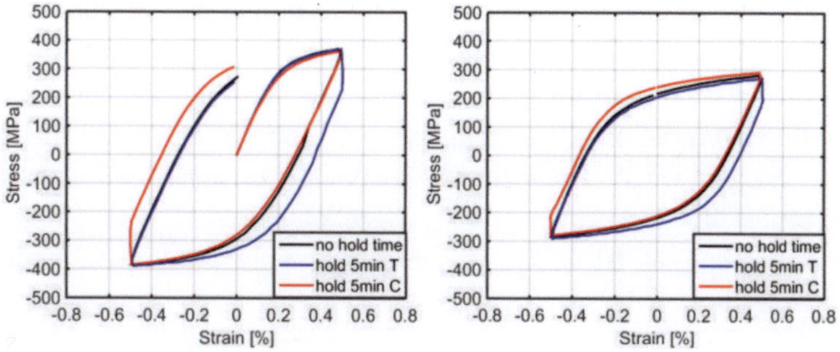

***a)** Hysteresis loops of the first cycle.* ***b)** Hysteresis loops of the cycle at the half lifetime.*

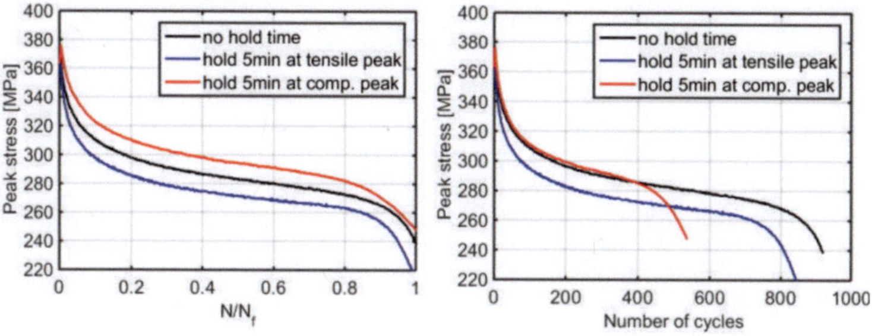

***c)** σ_{peak} vs. normalized number of cycles* ***d)** σ_{peak} vs. real number of cycles*

Fig. 5.6: Strain-controlled LCF tests at 550 °C performed with strain range 1.0% and various hold times.

As shown in Fig. 5.6, the peak tensile stress in the test performed with 5 min hold time at compressive peak was around 10 MPa larger than that without hold time, while the test performed with 5 min hold time at tensile peak was around 10 MPa smaller than that without hold time. This phenomenon explains why the lifetime with compressive hold is shorter than those without and with tensile hold, as shown in Fig. 5.6d. The lifetime with tensile hold is shorter than that without hold time, which is owing to the larger inelastic strain in each cycle, according to the Coffin–Manson relationship [52-54]. Since the result is comparable to those reported by Shankar et al. [82], in the sense that the lifetime with compressive hold is much shorter than that with tensile hold or without hold, the assistant effect of compressive hold time to oxide layer is also an important reason of reduced lifetime with compressive hold. However, the topic of fatigue damage and lifetime is not the focus of the current work. The material responses in these two tests with hold times illustrate the static recovery of kinematic hardening and cyclic softening of P91 at 550 °C, which are important for the modeling approach afterwards.

5.2 Results of Stress-controlled Ratcheting Tests

In the second stage of experiments at 550 °C, a variety of groups of uniaxial stress-controlled tests were performed. Fig. 5.7 illustrates one of the tests at 550°C performed with σ_{peak}= 325 MPa and σ_{mean}= 10 MPa. The stress rate is $\pm$50 MPa/s, which is the same for that in Fig. 4.6 in the previous chapter. As at RT, the stresses are in form of triangle wave (see Fig. 4.6 a) and strains are induced by stresses (see Fig. 4.6 b); the induced ε_{mean} increase with number of cycles, which shows ratcheting behavior (see Fig. 4.6e), and strain ranges also increase owing to cyclic softening (see Fig. 4.6 d).

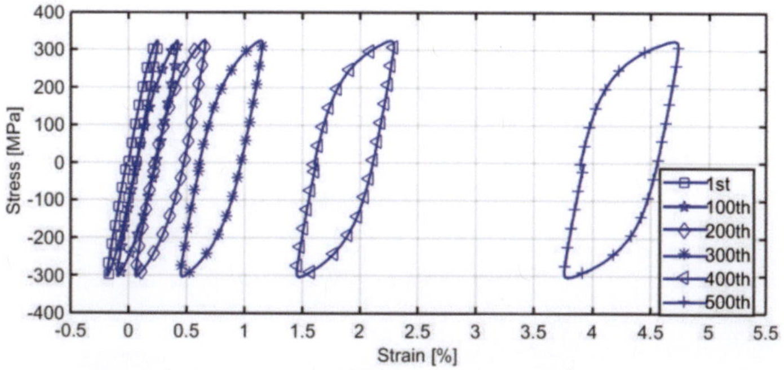

Fig. 5.7: Ratcheting tests at 550°C performed with σ_{peak} = 325 MPa, σ_{mean} = 10 MPa, stress rate ±50 MPa/s, hysteresis loops of the 1st 100th 200th 300th 400th 500th cycles.

In the following Sections 5.2.1~5.2.5, various groups of results are illustrated, corresponding to Table 3.5 and Table 3.6.

5.2.1 Influence of peak stress

According to Table 3.5a, six uniaxial ratcheting tests were performed to investigate the influence of peak stress. ε_r versus number of cycles is plotted in Fig. 5.8. Since the $\acute{\varepsilon}_r$ of the test performed with σ_{peak} = 275 MPa was negligible comparing to the other five tests, it is separately plotted in Fig. 5.8b. It is obvious that $\acute{\varepsilon}_r$ is higher with larger σ_{peak}.

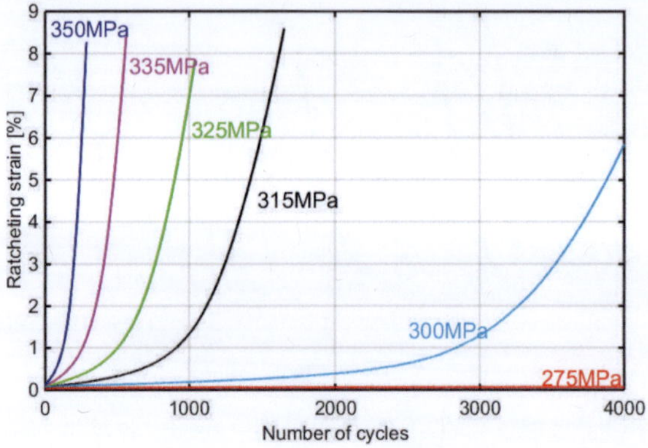

a)

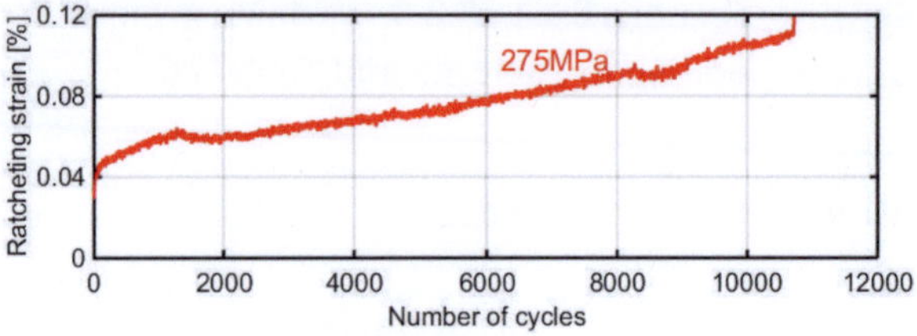

b)

Fig. 5.8: Ratcheting tests at 550 °C performed with σ_{mean} = 25 MPa, stress rate ±50 MPa/s, various σ_{peak}, ε_{r} vs. number of cycles.

$\Delta\varepsilon^{\text{in}}$ versus number of cycles of the six ratcheting tests is plotted in Fig. 5.9. The evolution of $\Delta\varepsilon^{\text{in}}$ is similar to that at RT (see Fig. 4.8), with an even and smooth increase of $\Delta\varepsilon^{\text{in}}$ owing to cyclic softening, however without showing two stages as in strain-controlled tests (see Fig. 5.3). Note that none of these six ratcheting tests end with macro cracks.

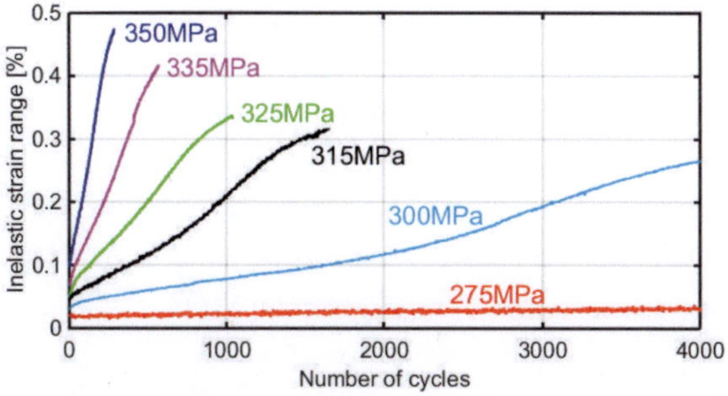

Fig. 5.9: Ratcheting tests at 550 °C performed with σ_{mean} = 25 MPa, stress rate ±50 MPa/s, various σ_{peak}, $\Delta\varepsilon^{in}$ vs. number of cycles.

When the $\acute{\varepsilon}_r$ are put together with their σ_{peak} in a log-linear plot (see Fig. 5.10), a quasi-linear relationship can be seen, which is similar to the case at RT (see Fig. 4.9). $\acute{\varepsilon}_r$ shown in Fig. 5.10 are the average rates until the cycle in which the maximum strain reaches 3% or cycle number achieves 10000, which ever comes first.

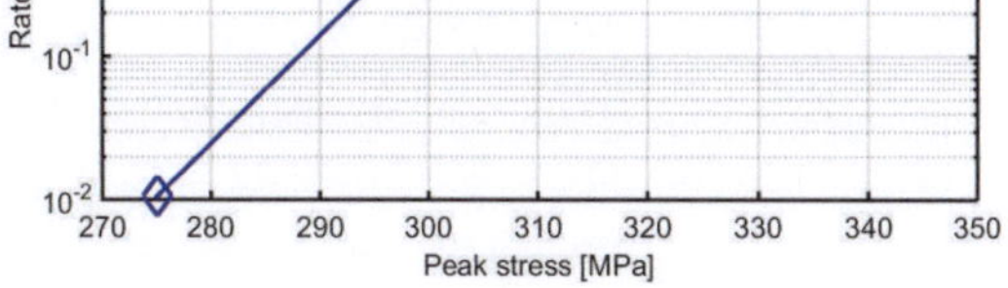

Fig. 5.10: Ratcheting tests at 550°C performed with σ_{mean} = 25 MPa, stress rate ±50 MPa/s, average ratcheting rates vs. σ_{peak} in log-linear diagram.

Similar to at RT, one more verification test was performed at 550 °C with σ_{max} = 310 MPa and σ_{min} = -325 MPa. As illustrated in Fig. 5.11, the absolute values of ε_r in the loading cases with stress ranges 325~-310 MPa and 310~-325 MPa agree with each other; in other words, the symmetric mean stress (±7.5 MPa) leads to symmetric ε_r. This symmetry can be compared with the asymmetry in the cases at RT reported in Section 4.2.1 (see Fig. 4.10), where the loadings with compressive σ_{mean} at RT lead to a similar but lower magnitude of $\dot{\varepsilon}_r$ than that with tensile σ_{mean}, if the absolute values of σ_{mean} are the same.

This symmetry shown in Fig. 5.11 supports the assumption that the material strength of P91 at 550 °C under tension and compression is the same.

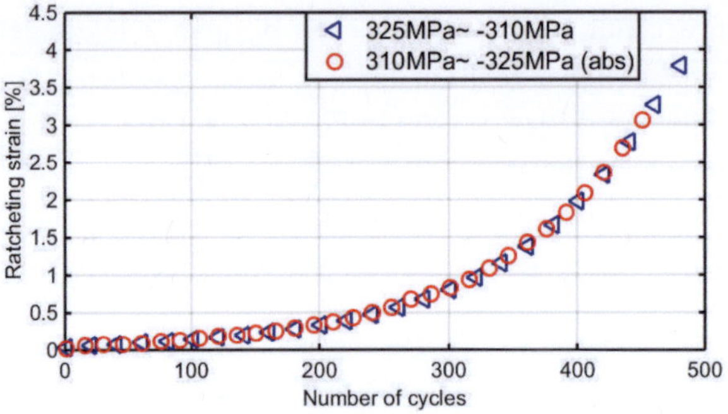

Fig. 5.11: A group of ratcheting tests at 550 °C performed with symmetric σ_{mean} (±7.5 MPa), ε_r vs. number of cycles.

5.2.2 Influence of Mean stress/Stress Ratio

According to Table 3.5b, 16 uniaxial ratcheting tests were carried out with the same maximum stress (peak tensile stress) of 325 MPa and various minimum stresses ranging from -335 MPa to zero. Fig. 5.12 illustrates all these ε_r versus number of cycles. Fig. 5.12b clarifies some cases with relatively lower $\dot{\varepsilon}_r$ in Fig. 5.12a while Fig. 5.12c further enlarges two cases in Fig. 5.12b.

The results are similar to the cases at RT in the sense that the maximum σ_{mean} (162.5 MPa) does not lead to fastest ratcheting with the same peak tensile stress. Instead, this maximum σ_{mean} leads to one of the two lowest $\acute{\varepsilon}_r$. The case with lowest $\acute{\varepsilon}_r$ is with σ_{mean} = 100 MPa, which is even slower than the case with σ_{mean} = 162.5 MPa (Fig. 5.12 c). It is clear that in the case with σ_{mean}= 162.5 MPa, the minimum stress is zero and the specimen is always under tensile loading. The creep strain leads to faster accumulated strain than the case with σ_{mean} = 100 MPa. After all, $\acute{\varepsilon}_r$ in these two cases is negligible compared to the rates in the other tests. It can be concluded that shake down happens in these two cases, which is similar to the case with a stress range of 500~0 MPa at RT, as shown in Fig. 4.14 in Section 4.2.2.

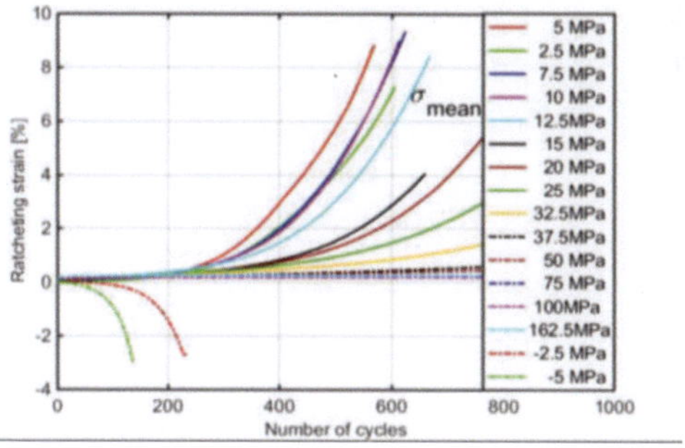

a)

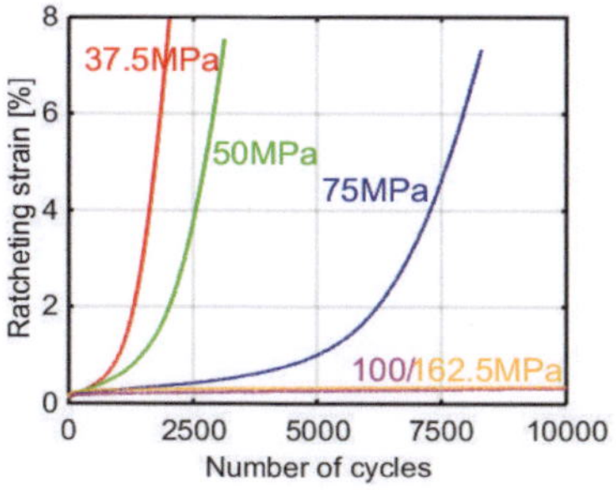

b)

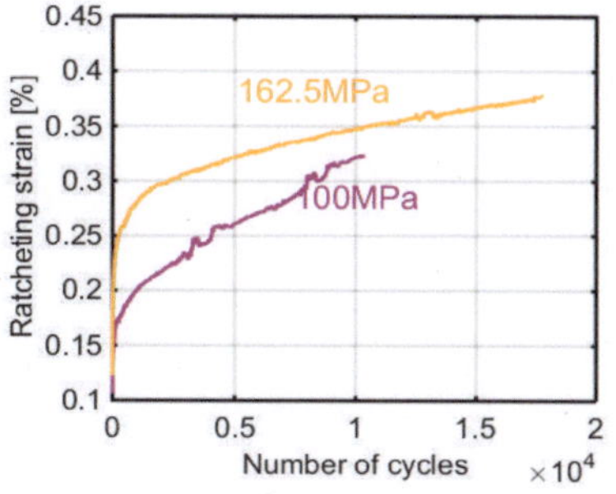

c)

Fig. 5.12: Ratcheting tests at 550 °C performed with σ_{peak} = 325 MPa, stress rate ±50 MPa/s, various σ_{mean}, ε_r vs. number of cycles.

$\Delta\varepsilon^{in}$ versus number of cycles of the 16 ratcheting tests are plotted in Fig. 5.13. It is clear that larger stress range leads to a higher increase in the rate of $\Delta\varepsilon^{in}$, similar to at RT (see Fig. 4.12) and the evolution of $\Delta\varepsilon^{in}$ owing to cyclic softening is also similar to that at RT. Still, no macro cracks were observed on any specimens of these 16 ratcheting tests.

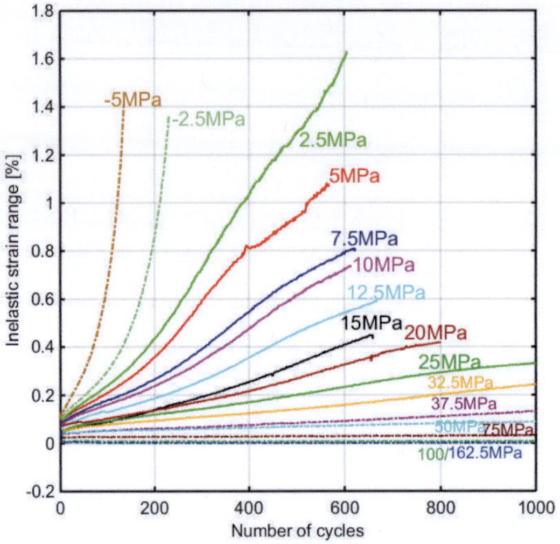

Fig. 5.13: Ratcheting tests at 550 °C performed with σ_{peak} = 325 MPa, stress rate ±50 MPa/s, various σ_{mean}, $\Delta\varepsilon^{in}$ vs. number of cycles.

Fig. 5.14 shows the relation between $\acute{\varepsilon}_r$ and stress ratios. The average $\acute{\varepsilon}_r$ together with corresponding stress ratios are plotted in this diagram. $\acute{\varepsilon}_r$ is calculated as the average rates until the cycle in which the maximum strain reaches ±3% or until 10000th cycle, whichever comes first. The maximum $\acute{\varepsilon}_r$ appears around a stress ratio of -0.95 and decreases with increasing stress ratio until 0. This phenomenon is similar to that at RT and the cause has already been discussed in Section 4.2.2.

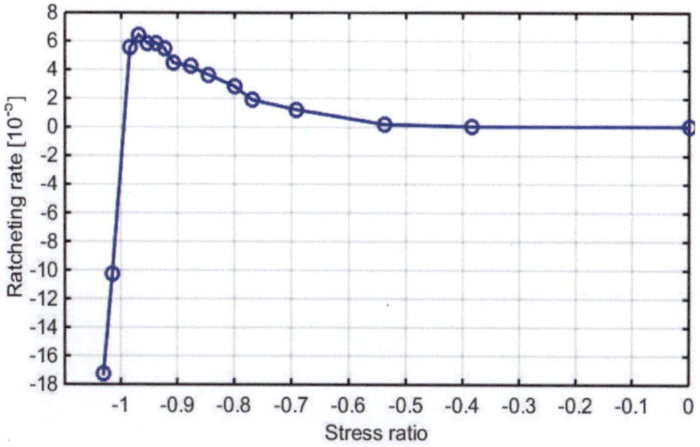

Fig. 5.14: Ratcheting tests at 550 °C performed with σ_{peak} = 325 MPa, stress rate ±50 MPa/s, average ratcheting rates vs. stress ratios.

5.2.3 Ratcheting with Zero Mean Stress

In the uniaxial symmetric strain-controlled LCF tests at 550 °C reported in Section 5.1, symmetric peak stresses are induced with multiple strain ranges, which indicates that the asymmetry of material strength under tension and compression at RT do not exist at 550 °C. Under stress-controlled loading, tests with zero σ_{mean} were performed to check whether positive ratcheting will happen, similar to that at RT.

As listed in Table 3.5c, four tests with zero σ_{mean} were performed and the results are plotted in Fig. 5.15. The results seem quite different from those at RT, which are reported in Section 4.2.3. At RT, symmetric stress-controlled loadings always lead to positive ratcheting and a larger σ_{a} leads to a higher positive $\varepsilon_{\mathrm{r}}'$.

However, in these four tests at 550 °C, stress amplitude σ_a = 350 MPa leads to positive ratcheting while the accumulated strains in tests performed with σ_a= 340 and 325 and 315 MPa have an uncertain direction. Further, all these four tests ended with fracture. Although the accumulated strains are not exactly zero, they reach no more than 0.5% until fracture, while in the cases at RT the strain can reach more than 5% without macro cracks. Therefore, the fracture comes much earlier than the problem of ratcheting at 550 °C. On the other hand, the accumulated strains in these tests are not exactly zero (due to non-ideally symmetric controlled loading) but are still negligible compared to those in the RT cases (see Section 4.2.3).

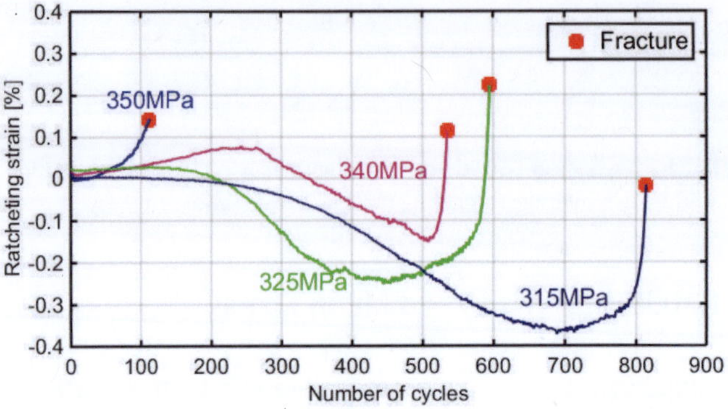

Fig. 5.15: Ratcheting tests at 550°C performed with zero σ_{mean}, stress rate ±50 MPa/s, various σ_a, ε_r vs. number of cycles.

$\Delta\varepsilon^{in}$ versus number of cycles is plotted in Fig. 5.17. All four tests ended with fractures, which led to a fast acceleration of increase of $\Delta\varepsilon^{in}$ at the end of the tests. However, except for the last acceleration, the evolution of $\Delta\varepsilon^{in}$ before macro cracks appeared is still similar to those in the other ratcheting tests at 550 °C under asymmetric loadings (see Fig. 5.9 and Fig. 5.13). Note that except for these four tests, all the other ratcheting tests at 550 °C ended without the appearance of macro cracks.

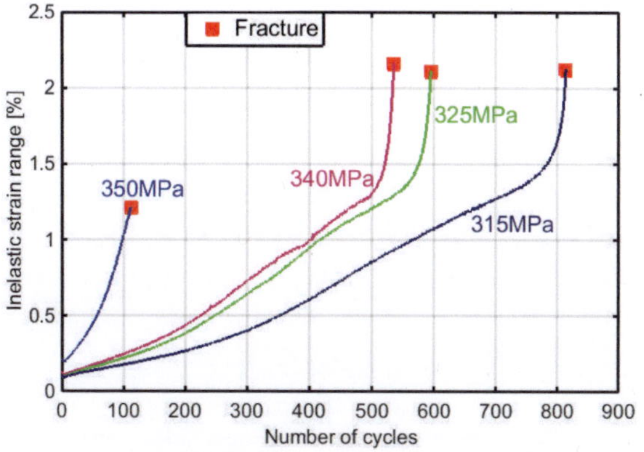

Fig. 5.16: Ratcheting tests at 550 °C performed with zero σ_{mean}, stress rate ±50 MPa/s, various σ_a, $\Delta\varepsilon^{in}$ vs. number of cycles.

On the contrary to the corresponding phenomena at RT, the material at 550 °C shows:

1. Symmetry of peak tensile and compressive stresses in strain-controlled LCF tests (see Fig. 5.5).
2. Identical magnitudes of $\acute{\varepsilon}_r$ with symmetric tensile/compressive σ_{mean} (see Fig. 5.11).
3. Negligible $\acute{\varepsilon}_r$ with zero σ_{mean} (see Fig. 5.15).

These three phenomenon together lead to the conclusion that the asymmetry of material strength under tension and compression at RT does not exist at 550 °C.

5.2.4 Influence of stress rate

According to Table 3.5d, two experiments are carried out were $\dot{\sigma}$ = ±10 and ±250 MPa/s to evaluate the visco-plasticity of P91 at 550 °C. Together with the previous test with $\dot{\sigma}$ = ±50 MPa/s, a diagram can be plotted as in Fig. 5.17. Compared to the results obtained at RT (see Section 4.2.4), it is clear that $\dot{\sigma}$ has larger influence on ratcheting behavior. According to the average $\acute{\varepsilon}_r$ before the peak strain reaches 3%, the test with ±10 MPa/s has twice higher $\acute{\varepsilon}_r$ than the test with ±50 MPa/s while in the same case at RT it is only 18.3% higher. The test with ±250 MPa/s has 26.5% lower $\acute{\varepsilon}_r$ than that with ±50 MPa/s at 550 °C, which is larger than the 22.3% obtained at RT. Hence, the material has lower viscosity at 550 °C than that at RT.

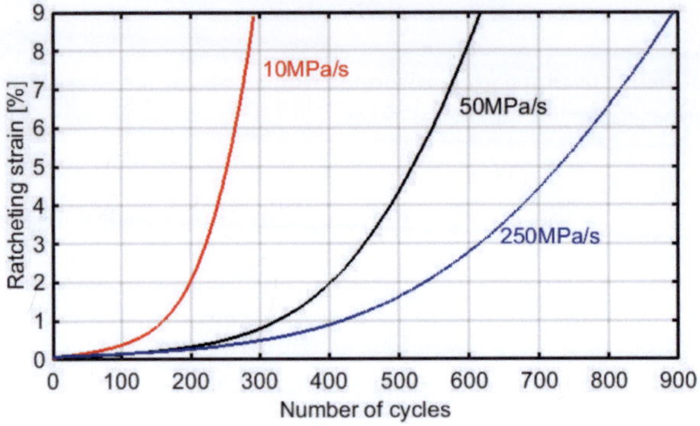

Fig. 5.17: Ratcheting tests at 550 °C performed with σ_{peak} = 325 MPa, σ_{mean} = 7.5 MPa, various stress rates, ε_r vs. number of cycles.

$\Delta\varepsilon^{in}$ versus number of cycles is plotted in Fig. 5.18. $\Delta\varepsilon^{in}$ in the test performed with stress rate ±10 MPa/s fluctuates, which was not observed in all other ratcheting tests at both temperatures. This could be owing to fluctuation of the oven temperature. In spite of this fluctuation of $\Delta\varepsilon^{in}$, the corresponding ε_r shows no influence by such fluctuation (see Fig. 5.17) and the evolutions of $\Delta\varepsilon^{in}$ in tests performed with $\dot{\sigma}$ = ±50 and ±250 MPa are similar to the other ratcheting tests at 550 °C under asymmetric loadings (see Fig. 5.9 and Fig. 5.13).

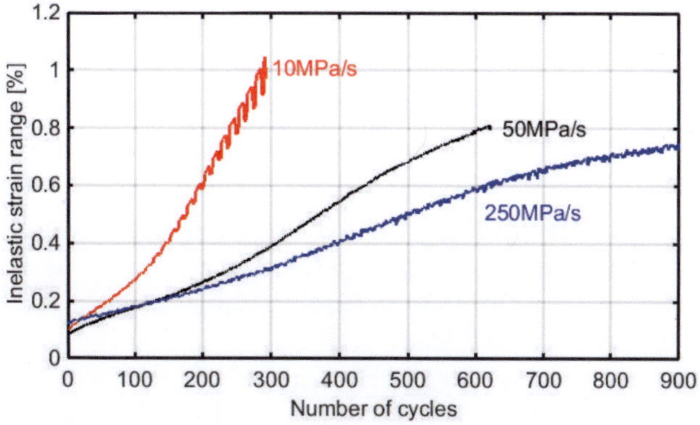

Fig. 5.18: Ratcheting tests at 550 °C performed with σ_{peak} = 325 MPa, σ_{mean} = 7.5 MPa, various stress rates, $\Delta\varepsilon^{\text{in}}$ vs. number of cycles.

5.2.5 Influence of Hold Time

Similar to at RT, several ratcheting tests were carried out with hold times at 550 °C, as listed in Table 3.5e. In addition, several creep tests were performed in comparison to the ratcheting tests with hold times according to Table 3.6.

Fig. 5.19 illustrates the above-mentioned experiments at 550 °C. The two ratcheting tests with 5 and 0.5 min hold times at tensile peaks are compared with two creep tests; one is the engineering-stress-controlled technical creep test, and the other one is the true-stress-controlled physical creep test, as shown in Fig. 5.19a. It is clear that the creep strain in the engineering-stress-controlled creep test is larger than that of the true-stress-controlled creep test. The accumulated strain in the ratcheting test performed with hold times of 5 min at tensile peaks (σ_{peak} = 325 MPa) is mainly composed of accumulated creep strains, and the accumulated strain is larger than the creep strain of σ_{creep} = 325 MPa in both engineering- and true-stress-controlled creep tests.

In Fig. 5.19b, the strain–time curves of two experiments are plotted: one is true-stress-controlled creep test with σ_{creep} = -310 MPa, and the other is ratcheting with hold times at compressive peaks σ_{min} = -310 MPa. It can be seen that the magnitude of this creep strain is lower than the magnitude of accumulated strain in the ratcheting performed with hold times at compressive peaks. Together with the comparison shown in Fig. 5.19a, we can come to conclusion that the cyclic loading accelerates the strain accumulation comparing to the corresponding pure creep test.

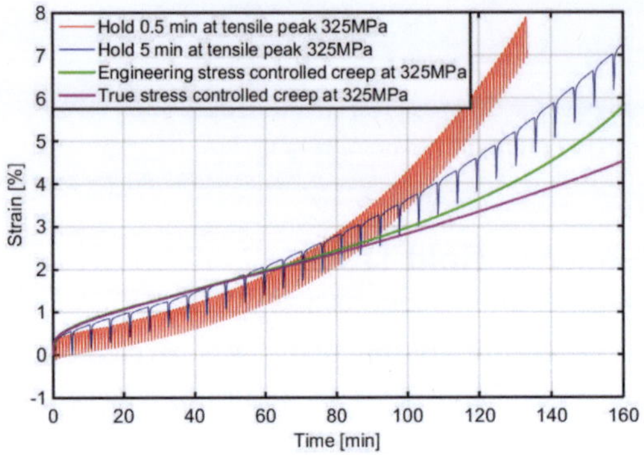

a) *Ratcheting tests performed with hold times of 0.5 & 5 min at peak tension, compared with eng.- & true-stress-controlled creep tests* σ_{creep} *= 325 MPa.*

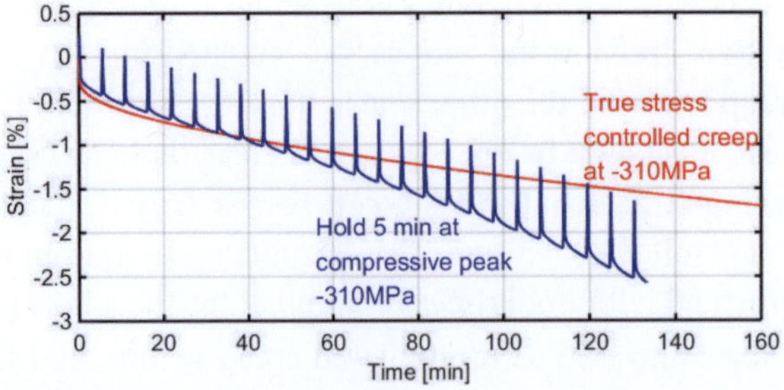

b) *Ratcheting tests performed with hold times of 5 min at peak compression, compared with true-stress-controlled creep test at* σ_{creep} *= -310 MPa.*

Fig. 5.19: Comparison between ratcheting tests with hold times and creep tests at 550 °C. σ_{peak} = 325 MPa, σ_{mean} = 7.5 MPa, stress rate ±50 MPa/s. Strain vs. time

The accumulated strain in the test performed with 0.5 min hold times at tensile peaks is firstly lower than the corresponding creep strain. However, it accelerates and surpasses the creep strain in the later stage. The reason is, on the one hand, the creep strains during hold times are increasing with increasing cycle number owing to softening, and on the other hand, the creep initially has a faster primary creep and then a slower secondary creep. The creep strain during each 0.5 min hold time in each cycle is plotted versus number of cycles in Fig. 5.20.

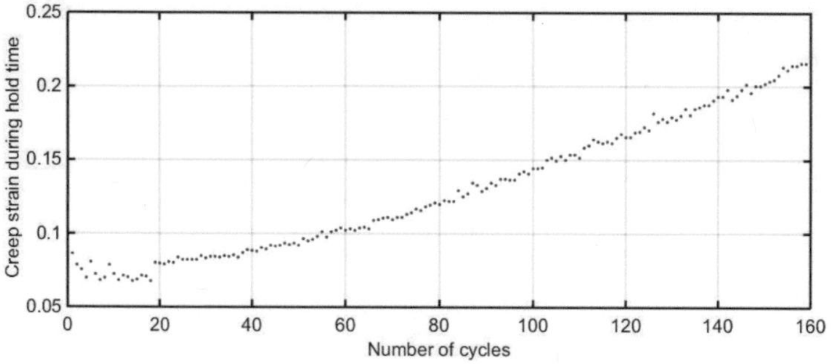

Fig. 5.20: Ratcheting tests at 550 °C performed with σ_{peak} = 325 MPa, σ_{mean} = 7.5 MPa, stress rate ±50 MPa/, hold 0.5 min at peak tension, creep strains during hold times vs. number of cycles.

These results at 550 °C can be compared with the results at RT with hold times at tensile peaks. As discussed in Section 4.2.5, in the test performed at RT with σ_{peak} = 500 MPa, σ_{mean} = 25 MPa and 10 min hold times at tensile peaks, the accumulated partial inelastic strain formed during the stress-changing process is negative (-15.63%), although σ_{mean} is tensile. Similar phenomena occur at 550 °C. For instance, in the case shown in Fig. 5.20, the total accumulated creep strains during hold times is 29.77% while the total strain is 8.34%, which means the accumulated partial inelastic strain formed during the stress-changing process is -21.43%, although σ_{mean} is tensile. The reason was discussed in

Section 4.2.5, that the material response to cyclic stress-controlled loading has a tendency to close the hysteresis loops.
However, with longer hold times at peak stress (i.e., 5 min), the material responses are more similar to those from the pure creep test. The total strain is mostly composed of accumulated creep strains during hold times. In addition, some more pure creep tests were performed according to Table 3.6. The result for σ = 310 MPa is the absolute value in the creep test with σ = -310 MPa. The results are plotted in Fig. 5.21, together with data obtained by Kimura et al. [106].

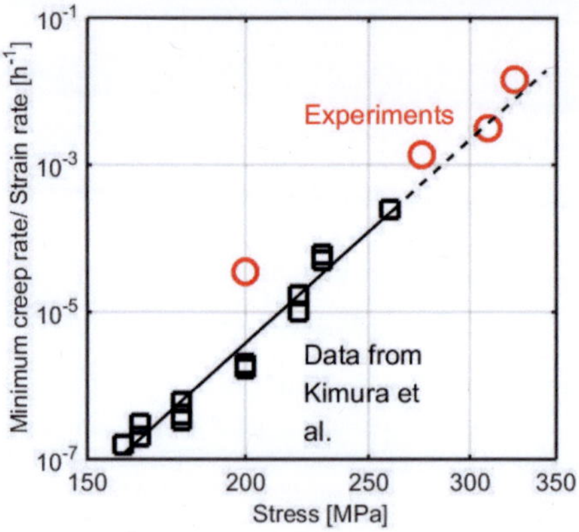

Fig. 5.21: Minimum creep rates vs. σ_{peak} for mod. 9Cr–1Mo steel at 550 °C.

The SCHENCK hydraulic testing machine in the current work was not specifically designed for, nor necessarily suitable for, creep tests. A special creep machine requires quite stable loading, usually controlled by gravity. However, in spite of the result with σ = 200 MPa, the minimum creep rates in the other three creep tests (marked as red circles) roughly agree with the extrapolation of the data obtained by Kimura et al. [106]. Hence, the results of the creep tests and ratcheting tests with hold times at 550 °C are reliable.

5.2.6 Diameter Check

After all experiments at 550 °C, the tested specimens were rechecked, because some specimens in 550 °C ratcheting tests showed obvious necking.

Detailed measurements were carried out to measure the diameters of tested specimens in different axial positions. The results for the high-temperature specimens are illustrated in Fig. 5.22.

As can be seen for various stress ranges, the minimum cross section always appears in the middle of the axial position. The only exception is the one with stress range 275~-225 MPa, where the accumulated strain is 0.11% only. In spite of this exception, the necking occurs and becomes more apparent if the strain is larger than 3%. In Fig. 5.22, the three diameters marked on each specimen include two near the end of the gage length and one in the middle. When the strain is larger than 3%, the difference in the cross section is above 5%, which is not acceptable.

Owing to the structural influence of necking, the measured data shows no longer pure material behavior. Therefore, only data before the total strain reaches 3% are taken to fit the value of material parameters in the modeling approach.

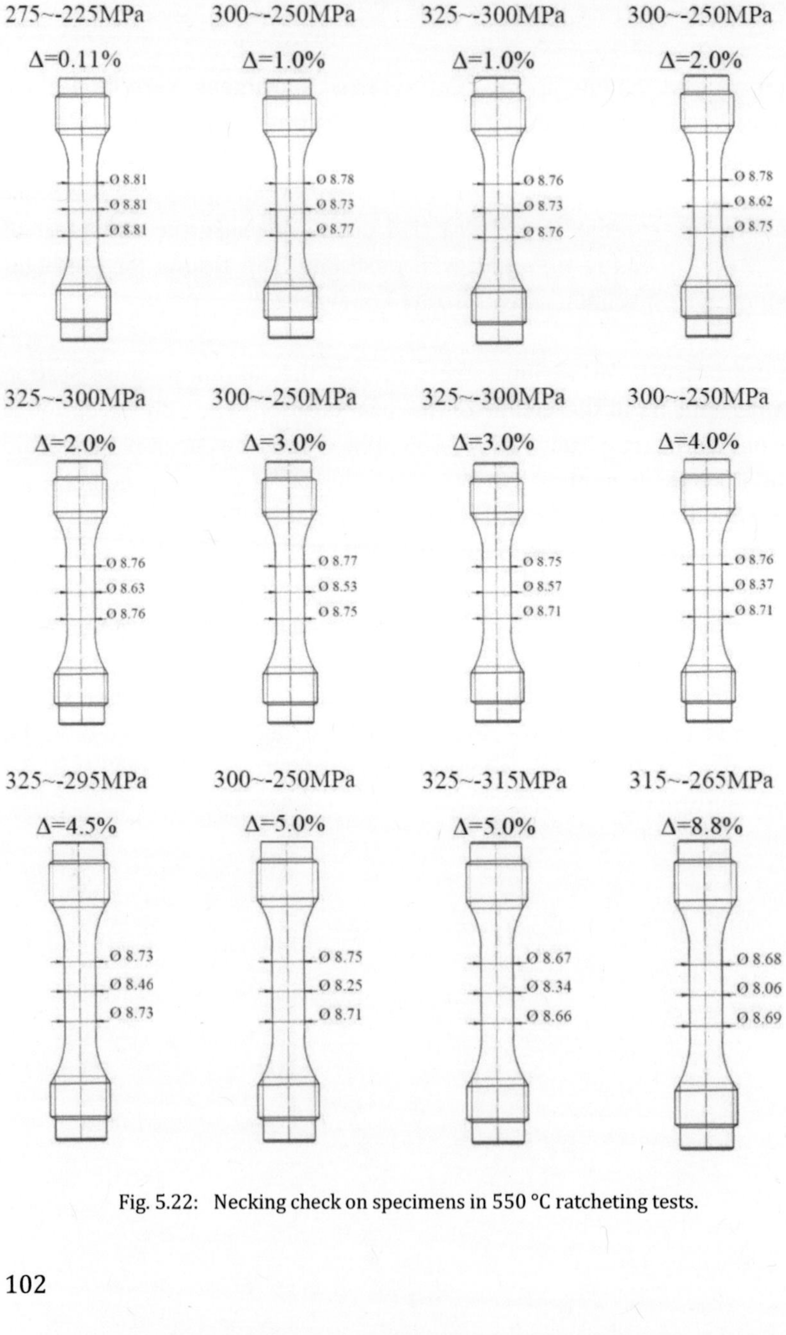

Fig. 5.22: Necking check on specimens in 550 °C ratcheting tests.

6 Simulation and Modeling

6.1 Requirements and Programming

Based on the experimental results reported in Chapters 1 and 1, a database including strain-controlled LCF tests and stress-controlled ratcheting tests for the P91 steel at both room temperature and 550 °C could be built. To analyze the mechanical behavior and predict material responses under more arbitrary loading conditions, a model was built to simulate the mechanical behavior of P91. The model description should agree with material responses qualitatively and quantitatively. According to experimental results at both temperatures, the simulated results should fulfill the following criteria:

1. Fitting on ε_r under multiple loading conditions.
2. Fitting on shapes of hysteresis loops.
3. Showing cyclic softening in strain-controlled LCF tests.
4. Larger σ_{peak} leads to higher $\acute{\varepsilon}_r$.
5. Highest $\acute{\varepsilon}_r$ with stress ratio of around -0.9~-0.95.
6. Strain range of hysteresis loop increases with increasing cyclic number.
7. Positive ratcheting with zero σ_{mean} at RT.
8. Higher $\dot{\sigma}$ leads to lower $\acute{\varepsilon}_r$ and vise versa.
9. Hold time at tensile and compressive peak leads to higher $\acute{\varepsilon}_r$ at RT.
10. Fitting on accumulation of strain in ratcheting tests at 550 °C performed with hold times and in creep tests

Owing to the complicated mechanical behavior and the simple geometry of tested specimens, analytical simulation was chosen, where a constitutive model could be built.

According to the review of work by a variety of scientists and engineers in Section 2.2, the constitutive model can originate from some current models and be modified to simulate the mechanical behavior of P91.

In the following discussion, a uniaxial constitutive model is firstly presented because all experiments reported in Chapter 1 and 1 are under uniaxial loading. Hence, the model in uniaxial form is enough to generate simulated results to compare with the performed experiments. Further extension to the multiaxial form is discussed afterwards.

The modeling approach is firstly applied on experiments at RT, and then on those at 550 °C.

The model under modification is Aktaa–Schmitt model [1]. This model is based on the so-called Chaboche model [5, 58-61]. which considers creep and plasticity as arising from the same dislocation source [62]. The Aktaa–Schmitt model has an additional description of cyclic softening, typically for Reduced Activation Ferritic Martensitic (RAFM) steels. Hence, this model is also known as the RAFM model. The equations of the RAFM model under modification are listed in Table 6.1 [1].

$\dot{\varepsilon} = \dot{\varepsilon}^{in} + \dot{\varepsilon}^{el}$	(6-1)
$\dot{\varepsilon}^{el} = \frac{\dot{\sigma}}{E}$	(6-2)
$\dot{\varepsilon}^{in} = \langle\frac{\lvert\Sigma\rvert - k}{Z}\rangle^{n}\, sgn(\Sigma)\ with\ \Sigma = \frac{\sigma}{\psi} - \Omega$	(6-3)
$\Omega = \Omega_1 + \Omega_2\ with\ \dot{\Omega}_i = H_i\dot{\varepsilon}^{in} - C_i\Omega_i\lvert\dot{\varepsilon}^{in}\rvert - R_i\lvert\Omega_i\rvert^{m_i-1}\Omega_i$	(6-4)
$\psi = \psi_1 + \psi_2, with\ \psi_1(t=0) = 0\ and\ \psi_2(t=0) = 1$	(6-5)
$\dot{\psi}_1 = -h\lvert\dot{\varepsilon}^{in}\rvert$	(6-6)

$\dot{\psi}_2 = c(\psi_s - \psi_2)\lvert\dot{\varepsilon}^{in}\rvert - r_\psi\lvert\psi_2 - \psi_r\rvert^{m_\psi - 1}(\psi_2 - \psi_r)$	(6-7)
$\psi_s = 1 - \psi_{s,\infty}(1 - exp(-c_s \max_{-\infty<\tau<t}\lvert\varepsilon^{in}(\tau)\rvert))$	(6-8)

Table 6.1: Constitutive equations in RAFM model

A difference between the model listed above and the original one in [1] is that the coupled damage is ignored in the current work because, except for one test at RT (σ_a = 520 MPa) and four tests at 550 °C (σ_a = 350, 340, 325, and 315 MPa) performed under symmetric loading, all other ratcheting tests reported in Chapter 1 and 1 were stopped before significant damage appeared: no kink on hysteresis loop was observed, such as the one shown in Fig. 4.1c, hence no macro cracks were observed on most specimens in the ratcheting tests. On the other hand, the coupled damage mentioned in [1] was for describing failure by fracture. From an engineering point of view, under asymmetrical stress-controlled cyclic loading, the failure of a structure owing to ratcheting should happen before the failure owing to fracture.

Another difference is in eq. (6-4) in which the back stress (BS) Ω is a superposition of more than one sub-BS. More sub-BSs lead to better fitting of the material response. In other modeling approaches, for example, the Ohno–Abdel–Karim model [6], the simulated BS is decomposed into eight components to fit the real material behavior. In the Yaguchi–Takahashi model [7], there are also eight sub-BS. However, it is supposed to minimize the number of BS components to simplify the simulation process and still give an acceptable simulation result. This can be seen in the later discussion. Note that the dynamic recovery of BS in eq. (6-4) follows the Armstrong–Frederick rule (see eq. (2-6)).

The static recovery is supposed to be thermally activated at generally high temperature. Therefore, to simplify the model for RT cases, parameter R_i in the term of static recovery of kinematic hardening (see eq. (6-4)) and r_ψ in the term of static recovery of isotropic softening (see eq. (6-7)) are set to be zero. ψ_s in eq. (6-7) is assumed to be constant as

$$\psi_s = 1 - \psi_{s,\infty}$$

for simplicity. Hence eqs. (6-4) and (6-7) are reduced into eqs. (6-9) and (6-10)

$$\dot{\Omega}_i = H_i \dot{\varepsilon}^{in} - C_i \Omega_i \left| \dot{\varepsilon}^{in} \right| \tag{6-9}$$

$$\dot{\psi}_2 = c\left(1 - \psi_{s,\infty} - \psi_2\right) \left| \dot{\varepsilon}^{in} \right| \tag{6-10}$$

Before experiments were carried out on P91, simulation programs had already been written in MATLAB. The programs were written according to the above-mentioned seven equations (eqs. (6-1), (6-2), (6-3), (6-5), (6-6), (6-9), (6-10)). Runge–Kutta fourth order iteration method is applied. Further, programs that can determine parameter values in these equations were written to fit the experimental data. The fitting programs are based on a fortran program MINUIT developed at CERN [107]. The main field of usage of MINUIT is statistical data analysis of experimental data recorded at CERN.

To verify the simulation ability and correctness of the calculation with MATLAB, programs were written according to the original RAFM model in [1] and with the same parameter values in [1] to check the model description compared to the material response of EUROFER97.
Fig. 6.1 illustrates the comparison between model and material responses for strain-controlled LCF tests on EUROFER97 at 550 °C. It is obvious that the simulation result is satisfactory, matching the shape of loops and decrease of peak tensile stresses (owing to cyclic softening). Hence, the calculation with MATLAB is proved to be correct.

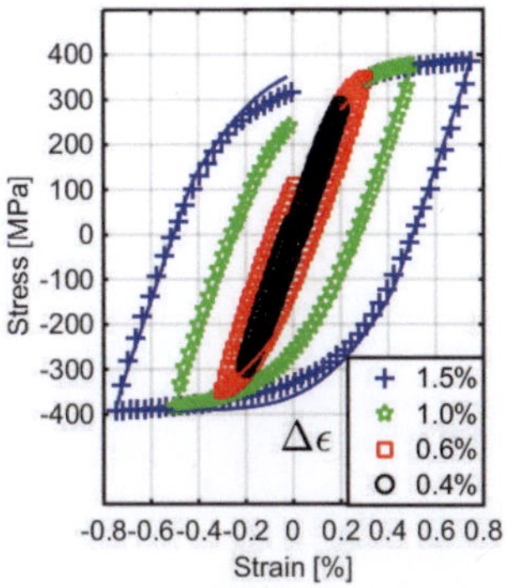

a) *hysteresis loops of the first cycles.*

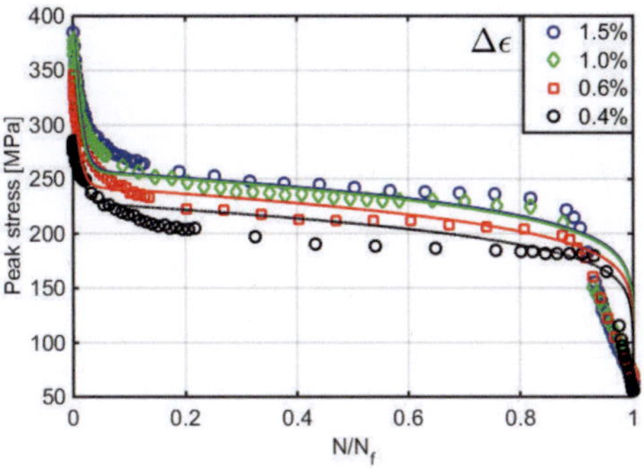

b) *peak tensile stresses vs. normalized number of cycles*

Fig. 6.1: Comparison between material response (markers) and model description (curves): Strain-controlled LCF tests on EUROFER97 at 550 °C.

In the iteration method, one of the key factors influencing the result is the step size. By calculating σ_{peak} in the cycle at half lifetime in the LCF test of EUROFER97 performed with strain range of 1.0% (marked in green in Fig. 6.1), different time steps were tested, as listed in Table 6.2. Except for the results with 0.1 and 0.09 seconds as time steps, all the other calculated σ_{peak} at half lifetime are almost equal. Hence, the iteration result converges with decreasing time step.

Time step [Second]	Iteration steps per cycle	σ_{peak} of half lifetime [MPa]
0.1	67	223.6
0.09	74	225.0
0.08	83	235.7
0.07	95	236.3987
0.05	133	236.4038
0.01	667	236.3985
0.001	6667	236.3986

Table 6.2: Test results with various time steps in Runge–Kutta fourth order iteration method.

In the simulation program in MATLAB, an automatic time-stepping algorithms was firstly defined in the following way: the iteration result $x(\Delta t_i)$ was calculated with time step Δt_i and $x(\Delta t_{i+1})$ was calculated with time step $\Delta t_{i+1} = \frac{\Delta t_i}{2}$. The initial time step $\Delta t_0 = 0.02s$. The iteration result $x(\Delta t_i)$ is reliable when

$$abs\left\{\frac{x(\Delta t_i) - x(\Delta t_{i+1})}{x(\Delta t_i)}\right\} < 10^{-6} \tag{6-11}$$

However, this time-stepping algorithm is time-costing during the simulation. A practical time-stepping algorithm is defined with the time-step chosen according to the increase/decrease rate of inelastic strain and making sure there are at least 500 iteration steps in each cycle:

$$\Delta t_i = \frac{\Delta t_0}{1 + abs\left|\Delta\varepsilon_{i-1}^{in}\Delta t_0/\Delta\varepsilon_0 \Delta t_{i-1}\right|} \tag{6-12}$$

in which $\Delta t_0 = 0.02$ s, $\Delta\varepsilon_0 = 6 \times 10^{-6}$, $\Delta\varepsilon_{i-1}^{in}/\Delta t_{i-1}$ is the increase/decrease rate of inelastic strain in the previous iteration. Function $\lceil x \rceil$ is the ceiling function, which is the smallest integer not less than x. Such a practical time-stepping algorithm leads to much shorter computation time.

The error owing to this practical time-stepping algorithm in this Runge–Kutta fourth order iteration method has been proved to be negligible, such as the difference between the results with time step 0.01 and 0.001 second in Table 6.2.

6.2 Initial Fitting of Parameter Values

As discussed in the previous section, simulation programs were written in MATLAB, according to the constitutive model constructed with seven equations (eqs. (6-1), (6-2), (6-3), (6-5), (6-6), (6-9), (6-10)). The next step is to fit parameter values for P91.

After strain-controlled LCF tests are finished at RT, the parameter values are fitted according to the material responses. The fitting process is as follows: E is the Young's modulus, which is calculated according to the slope of stress–strain unloading line in the first cycle. Parameters k, Z , n, H_1, C_1, H_2, and C_2 are fitted to make each simulated stresses agree with every measured data point in the first cycle.

The fitting algorithm mixes the sum of absolute relative errors between experimental results x_{test} and simulated results x_{sim} (eq. (6-13)) and the sum of absolute errors (eq. (6-14)). The calculation of these errors has less calculation time in MATLAB than using, for example, the least-square method.

$$\Delta_1 = \sum_{i=1}^{N} \left\{ \left| 1 - \frac{x_{test,i}}{x_{sim,i}} \right| + \left| 1 - \frac{x_{sim,i}}{x_{test,i}} \right| \right\} \tag{6-13}$$

$$\Delta_2 = \sum_{i=1}^{N} \left| x_{test,i} - x_{sim,i} \right| \tag{6-14}$$

The sum of absolute errors is used when the stress–strain hysteresis loops are fitted. The sum of absolute relative errors is used when the peak stresses are fitted. Applying the sum of absolute relative errors has better performance than, for example, the least-square method when the magnitudes of experimental results, for example peak stresses, range from relatively low to very high values.

As shown in Fig. 6.2a, the simulated hysteresis loops agree with experimental ones in the first cycle for various total strain ranges.

Afterwards parameters h, c and $\psi_{s,\infty}$ are fitted to have the simulated σ_{peak} match the experimental σ_{peak} in each cycle till the end of saturation stage of cyclic softening. As shown in Fig. 6.2c, the σ_{peak} in the simulated results shows a fast decrease in the first stage and a slow decrease in the second stage (saturation stage), which is the same as in material responses. Fig. 6.2b illustrates the hysteresis loops of the cycles at the half N_0. It is clear that the model description qualitatively and quantitatively agrees with the material responses in the strain-controlled LCF tests on P91 at RT.

N_0 is a roughly defined cycle number until which the data are considered to be reliable, as discussed in Section 4.1.

The 11 fitted parameter values are listed in Table 6.3.

$E\ (MPa)$	220086
$k\ (MPa)$	142.8
$Z\ (MPa\ s^{1/n})$	55.259
n	151.38
$H_1\ (MPa)$	460394
C_1	1979
$H_2\ (MPa)$	44984
C_2	256.7
h	1.333×10^{-3}
c	0.51099
$\psi_{s,\infty}$	0.1077

Table 6.3: Parameter values determined for P91 at RT based on strain-controlled LCF tests.

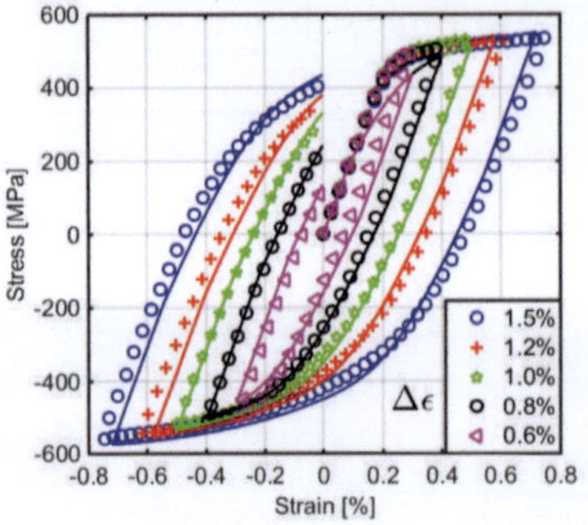

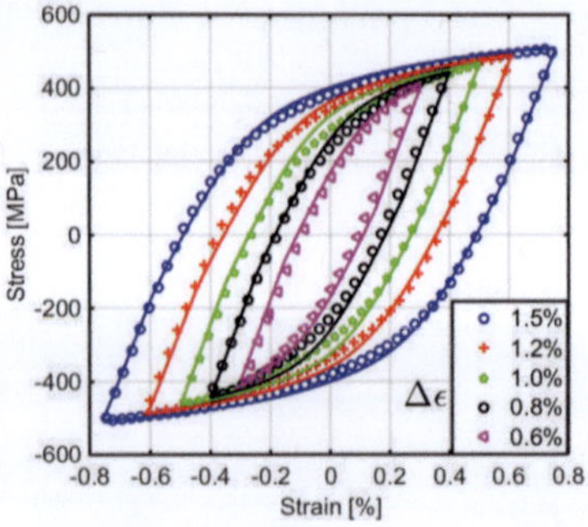

***a)** Hysteresis loops of the first cycles* ***b)** Hysteresis loops of the cycles at $N_0/2$*

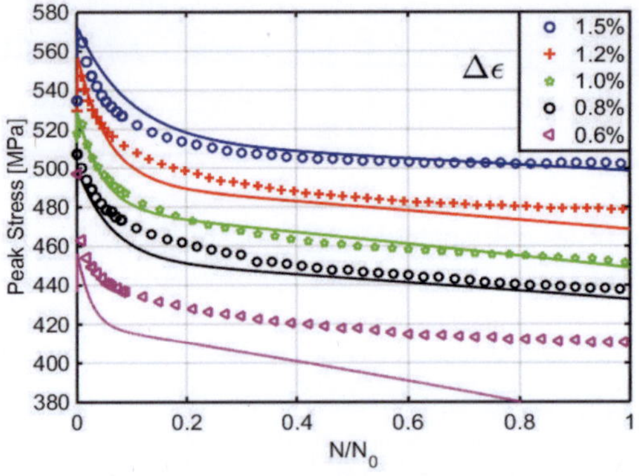

***c)** Peak tensile stresses vs. normalized number of cycles.*

Fig. 6.2: Comparison between material response (markers) and RAFM model description (curves): Strain-controlled LCF tests on P91 at RT performed with various strain ranges.

However, with the same model and the same parameter values, a stress-controlled test performed with σ_{peak} = 500 MPa and σ_{mean} = 25 MPa is simulated. Fig. 6.3 illustrates the comparison between the material response and model description. Apparently, the accumulated strain in the simulation is much larger than that in material response, although the shape of the simulated hysteresis loop in the first cycle is similar to the experimental one. Therefore, the RAFM model with Armstrong–Frederick rule for kinematic hardening dramatically overestimates $\acute{\varepsilon}_r$.

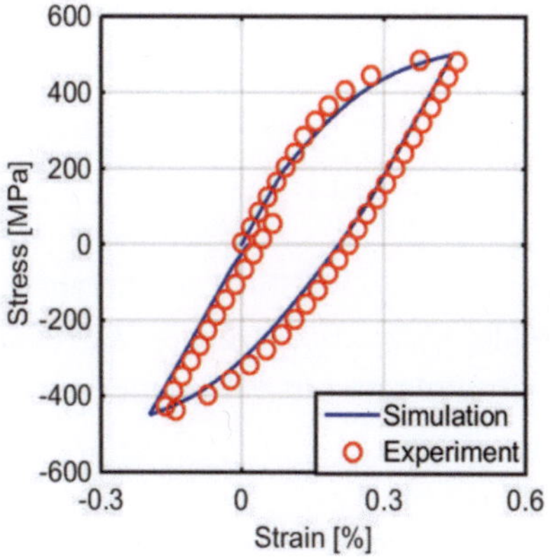

a) *loop of the first cycle*

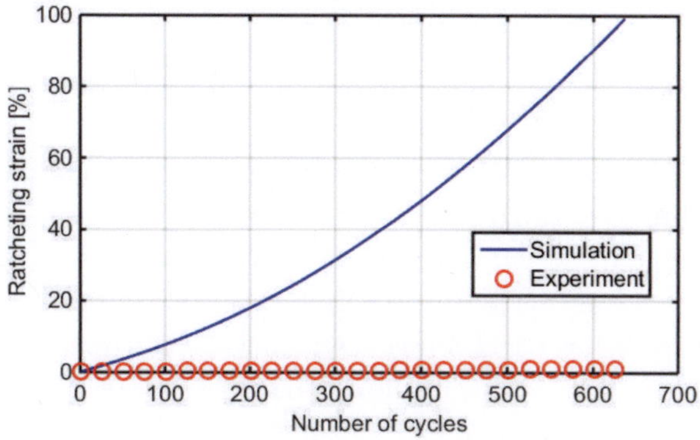

b) *accumulation of strain vs number of cycles.*

Fig. 6.3: Comparison between material response (markers) and RAFM model description (curves): Stress-controlled test at RT performed with σ_{peak} = 500 MPa, σ_{mean} = 25 MPa, stress rate ±50 MPa/s. Left: loop of the first cycle.

In the aim of fitting ε_r under multiple loading conditions, which is the first criterion for modeling listed at the beginnning of Section 6.1, the current RAFM model is modified, as discussed in the following sections.

6.3 Testing with Existing Models

6.3.1 Testing with RAFM Model

As illustrated in the previous section, the current RAFM model shows good ability to simulate strain-controlled LCF tests, but it obviously overestimates the accumulated strain under stress-controlled cyclic loading.

Different parameter values were tested in the current constitutive model (eqs. (6-1), (6-2), (6-3), (6-5), (6-6), (6-9), (6-10)) to simulate ratcheting. It was found that, when the parameters C_1 or C_2 are set to be zero, the model can yield no more ratcheting, which means if any one of the BS components has no dynamic recovery, there is no more ratcheting in the simulation.

The simulated results with different values of C_1 and C_2 are illustrated in Fig. 6.4. As can be seen in Fig. 6.4a, if the value of C_1 is kept as the value determined in the initial fitting of parameter values (shown in Table 6.3), the simulated hysteresis loop in the first cycle still approximately agrees with the experimental data. However, ε_r in the simulated result is, comparing to material response, negligible, because $C_2 = 0$. If the value of C_1 is set to be zero, as shown in Fig. 6.4b and c, even the simulated hysteresis loop in the first cycle is far from acceptable, with straight lines and sharp corners. The reason is that the parameters H_i and C_i of BS component Ω_1 are much larger than the parameters of Ω_2 (shown in Table 6.3), consequently Ω_1 determines the shapes of the hysteresis loops (matching the plastic modulus) near the yield point, while Ω_2 can make the simulated loops much closer to the experimental loops at higher strain, but

is still much smaller than Ω_1 and has a much smaller influence on simulated loop shapes.
Beside the influence on loop shapes, as long as one of the dynamic recovery terms is set to be zero, the accumulated strain in model description is negligible compared to material response.

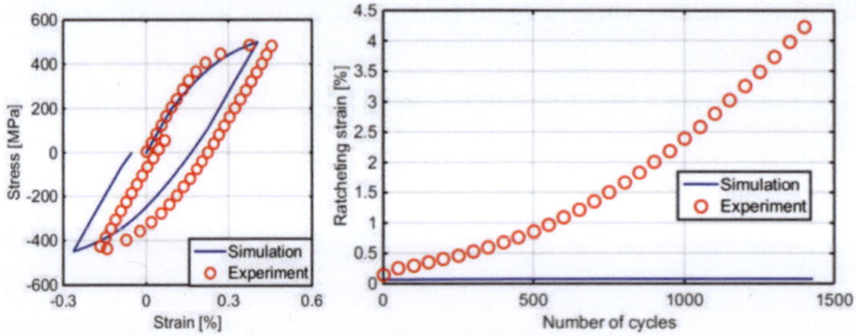

a) $C_1=1979, C_2=0$

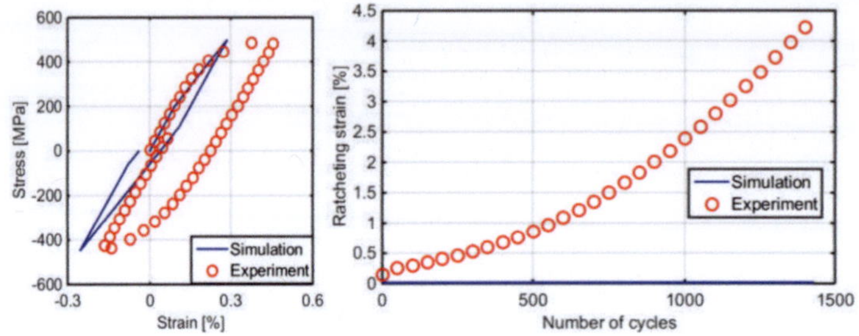

b) $C_1=0, C_2=0$

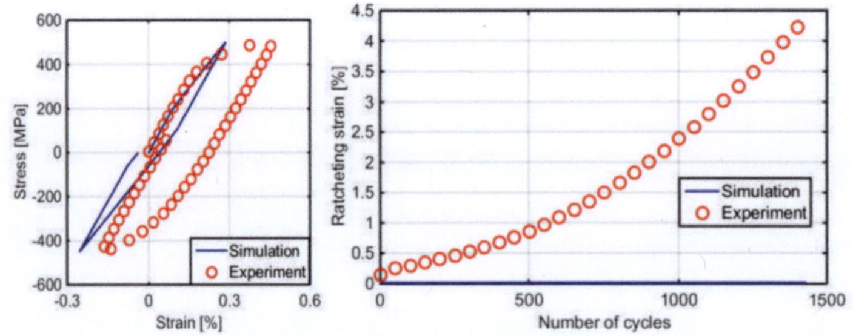

c) $C_1=0, C_2=256.7$

Fig. 6.4: Comparison between material response (markers) and RAFM model description (curves) with various parameter values for dynamic recovery: Stress-controlled tests at RT performed with σpeak = 500 MPa, σmean = 25 MPa, stress rate $\pm$50 MPa/s. Left: loop of the first cycle. Right: accumulation of strain vs number of cycles.

To fit ε_r in the case with σ_{peak} = 500 MPa, σ_{mean} = 25 MPa, and $\dot{\sigma}$ = $\pm$50 MPa/s at RT, the parameter C_2 is set to be 5, then both the loop shape and accumulated strain in the simulated results match the experimental data satisfactorily, as shown in Fig. 6.5a.

However, with the same parameter values (C_1 = 1979, C_2 = 5), the RAFM model description for the case with σ_{peak} = 500 MPa, σ_{mean} = 250 MPa, and $\dot{\sigma}$ = $\pm$50 MPa/s at RT has a great difference from the corresponding material response, as shown in Fig. 6.5b. Hence, the value of C_2 is not constant, but a function of the loading, which indicates that the main task to fit ε_r under multiple loading conditions is to modify the term for dynamic recovery of BS 2.

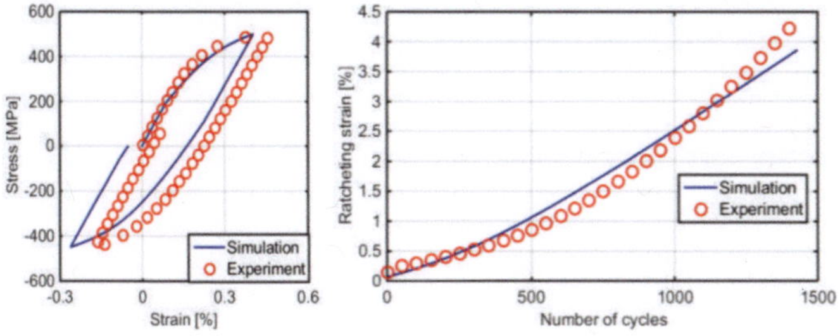

a) *σ_{peak} = 500 MPa, σ_{mean} = 25 MPa, stress rate $\pm$50 MPa/s at RT*

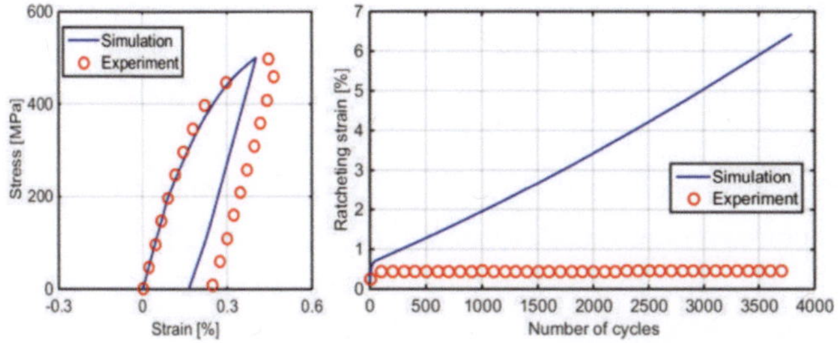

b) *σ_{peak} = 500 MPa, σ_{mean} = 250 MPa, stress rate $\pm$50 MPa/s at RT*

Fig. 6.5: Comparison between material response (markers) and RAFM model description (curves): Stress-controlled tests with C_1= 1979 and C_2= 5. Left: loop of the first cycle. Right: accumulation of strain vs. number of cycles.

6.3.2 Testing with Ohno-Wang Model I (OW I)

In the modeling approaches reported in [3, 4, 6, 30, 51, 65, 66], the main attention was paid to modification of the term of dynamic recovery of the kinematic hardening.

As mentioned in review of Ohno and Wang [3, 4] in Section 2.2, a critical value r_i for the magnitude of BS component Ω_i was assumed in OW I: If the magnitude of Ω_i reaches the value of r_i, the magnitude of Ω_i stops increasing, as illustrated in Fig. 2.6.

According to [3], eq. (6-9) is altered into:

$$\dot{\Omega}_i = H_i\dot{\varepsilon}^{in} - Hs({\Omega_i}^2 - \left(\frac{H_i}{C_i}\right)^2)C_i\Omega_i\langle\dot{\varepsilon}^{in}\frac{\Omega_i}{|\Omega_i|}\rangle \quad (6\text{-}15)$$

in which the dynamic recovery of kinematic hardening is only activated when the magnitude of Ω_i reaches the critical value of $\frac{H_i}{C_i}$. Hs denotes the Heaviside step function as mentioned in Section 2.2. On the other hand, the magnitude of Ω_i will stay constant after it reaches the critical value, as reviewed in Section 2.2.

By changing both equations for Ω_1 and Ω_2 in to eq. (6-15), the results are shown in Fig. 6.6a. If the equation of Ω_1 is kept as eq. (6-9) and only the equation of Ω_2 is changed to eq. (6-15), the results are shown in Fig. 6.6b.

Comparing the results shown in Fig. 6.6a and b, it is clear that if both equations of dynamic recovery of Ω_1 and Ω_2 are altered into the form of OW I (eq. (6-15)), the simulated hysteresis loops are composed of straight lines and sharp corners, while keeping the equation of Ω_1 as in Armstrong–Frederick rule (eq. (6-9)), the simulated loop shape of the first cycle is closer to the experimental data. However, both cases in Fig. 6.6 dramatically underestimate ε_r. As discussed in [3, 4], OW I yields no ratcheting.

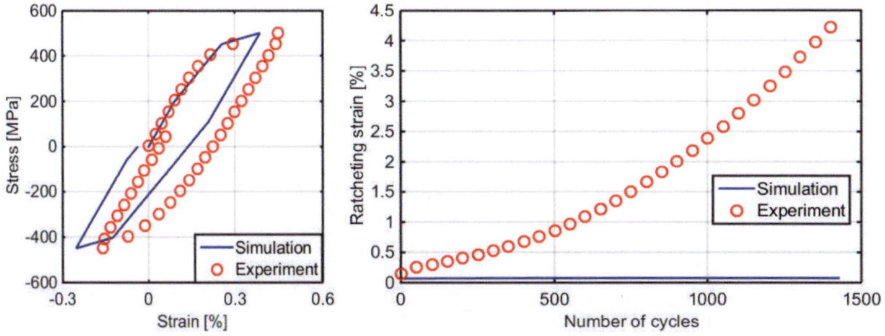

a) *Both Ω_1 & Ω_2 follows OW I rule.*

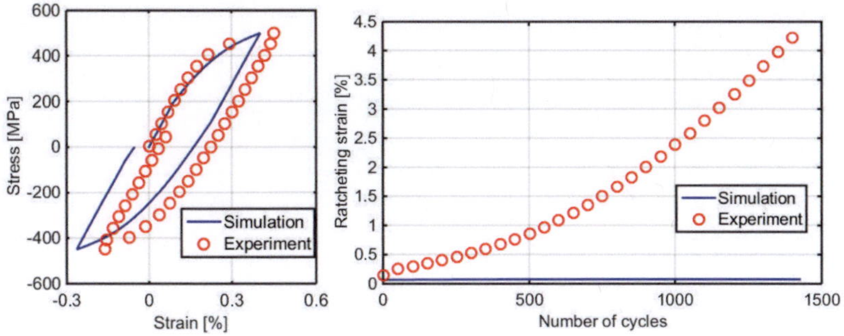

b) *Ω_1 follows AF rule and Ω_2 follows OW I rule.*

Fig. 6.6: Comparison between material response (markers) and OW I description (curves): Stress-controlled tests at RT performed with σ_{peak} = 500 MPa, σ_{mean} = 25 MPa, stress rate 50 MPa/s. C_1 = 1979, C_2 = 256.7, Left: loop of the first cycle. Right: accumulation of strain vs. number of cycles.

In [3, 4], eight instead of two components of BS were assumed to make simulated loop shapes closer to those in the experimental results; in other words, shorter and denser straight lines in the simulation to approximate the smooth curves in the material response. Still, the simulated loops were piecewise linear and expressed no uniaxial ε_r [3, 4].

6.3.3 Testing with Ohno–Wang Model II (OW II)

To avoid the multi-linear behavior in the model description of OW I, OW II was developed in which the Heaviside function $Hs({\Omega_i}^2 - \left(\frac{H_i}{C_i}\right)^2)$ was replaced by the term $\left(\frac{\Omega_i C_i}{H_i}\right)^{m_i}$. The BS equation is as follows:

$$\dot{\Omega}_i = H_i \dot{\varepsilon}^{in} - \left(\frac{\Omega_i C_i}{H_i}\right)^{m_i} C_i \Omega_i \left\langle \dot{\varepsilon}^{in} \frac{\Omega_i}{|\Omega_i|} \right\rangle \tag{6-16}$$

Eq. (6-16) reduces to eq. (6-15) when $m_i \rightarrow \infty$.

Fig. 6.7 illustrates the simulated results with application of OW II.

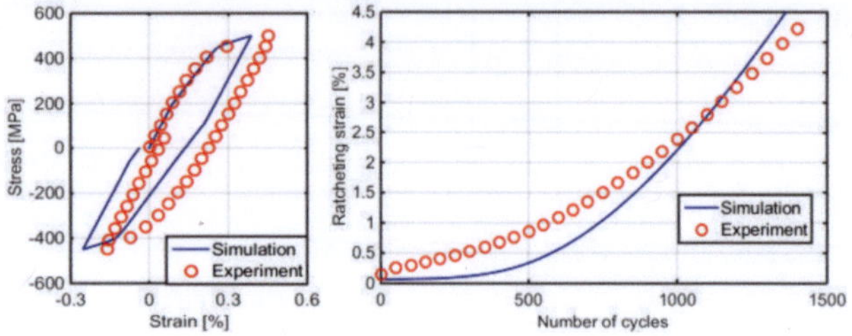

a) *Both Ω_1 & Ω_2 follows OW II rule.*

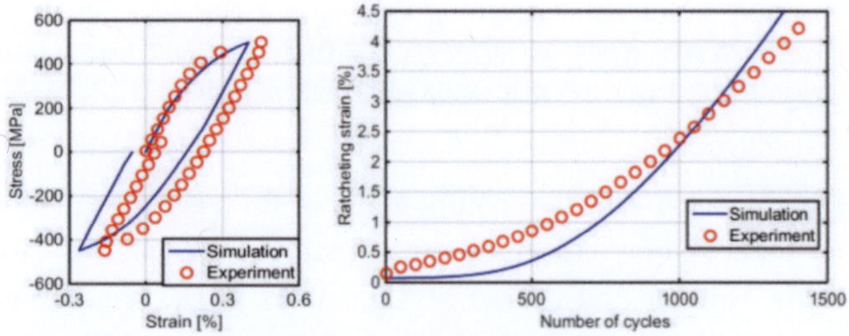

b) *Ω_1 follows AF rule and Ω_2 follows OW II rule.*

Fig. 6.7: Comparison between material response (markers) and OW II description (curves): Stress-controlled tests at RT performed with σ_{peak} = 500 MPa, σ_{mean} = 25 MPa, stress rate 50 MPa/s. C_1 = 1979, C_2 = 256.7 . Left: loop of the first cycle. Right: accumulation of strain vs. number of cycles.

The equations of Ω_1 and Ω_2 are changed into eq. (6-16), and suitable values of m_1 and m_2 are searched. It is found that if $m_1 > 8$ and $m_2 = 7$, the model yields a similar ε_r as the material and the value of m_1 does not affect the simulated ε_r much as long as $m_1 > 8$. By setting $m_1 = 8$ and $m_2 = 7$, the loop of the first cycle and accumulation of strain are shown in Fig. 6.7a. The loop is, as in Fig. 6.7 a), piecewise linear. If the equation of Ω_1 is kept as eq. (6-9) and we change only the equation of Ω_2 into eq. (6-16), the loop can be much smoother, as shown in Fig. 6.7b, and the simulated ε_r is similar to that shown in Fig. 6.7a. Further, similar to the case with σ_{mean} = 25 MPa, the model and material response in the case with σ_{peak} = 500 MPa, σ_{mean} = 250 MPa, and $\dot{\sigma} = \pm 50$ MPa/s is compared, as shown in Fig. 6.8. Parameter values are the same as the cases shown in Fig. 6.7.

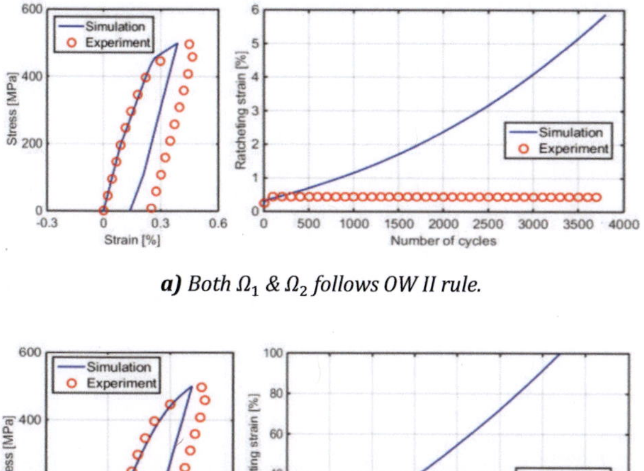

a) *Both Ω_1 & Ω_2 follows OW II rule.*

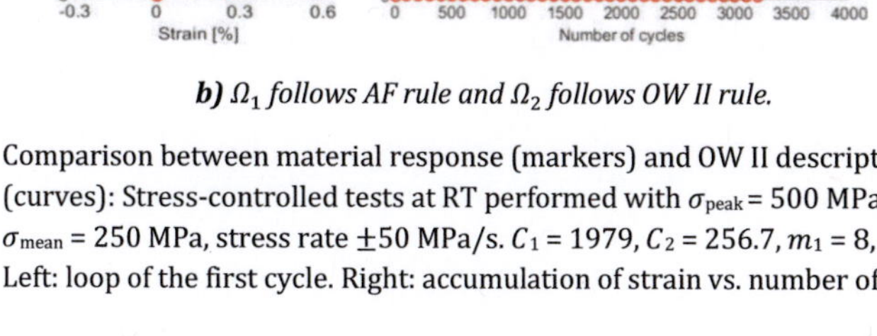

b) *Ω_1 follows AF rule and Ω_2 follows OW II rule.*

Fig. 6.8: Comparison between material response (markers) and OW II description (curves): Stress-controlled tests at RT performed with σ_{peak} = 500 MPa, σ_{mean} = 250 MPa, stress rate ± 50 MPa/s. $C_1 = 1979$, $C_2 = 256.7$, $m_1 = 8$, $m_2 = 7$, Left: loop of the first cycle. Right: accumulation of strain vs. number of cycles.

Fig. 6.8a shows the comparison with both equations for Ω_1 and Ω_2 changed to eq. (6-16), while Fig. 6.8b shows the comparison with only the equation for Ω_2 changed to eq. (6-16). As can be seen, OW II overestimates the ε_r with larger σ_{mean}, such as in this case with σ_{mean} = 250 MPa, the model yields too much ε_r compared to the material response.

Further, another case was checked in which hold times at tensile peaks are included. The comparison between model description and material response is illustrated in Fig. 6.9. All parameter values and equations are the same as the cases shown in Fig. 6.7a and Fig. 6.8a. As can be seen, OW II dramatically underestimates the accumulated strain for the case with hold time at peak tensile stress. Note that the simulated creep during hold time at tensile peak in the case shown in Fig. 6.9 is too small to be seen in the current view, while the creep in experiment is obvious.

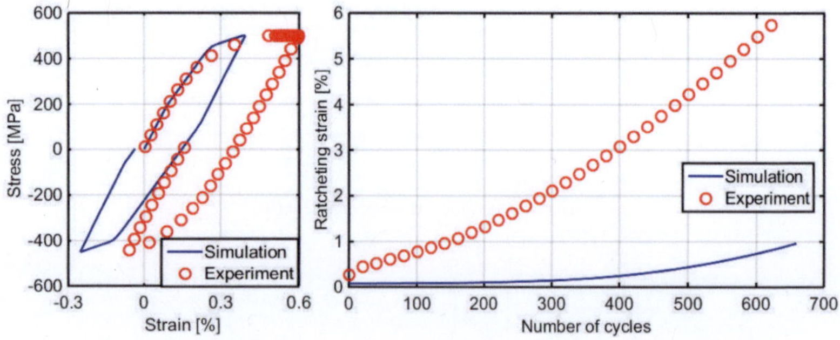

Fig. 6.9: Comparison between material response (markers) and OW II description (curves): Stress-controlled tests at RT performed with σ_{peak} = 500 MPa, σ_{mean} = 25 MPa, stress rate ±50 MPa/s, and hold 10 min at tensile peak. C_1 = 1979, C_2 = 256.7, m_1 = 8, and m_2 = 7. Left: loop of the first cycle. Right: accumulation of strain vs. number of cycles.

6.4 New Model for Room Temperature

Both the Ohno–Abdel–Karim model [6] and the Kang model [30, 51, 65] (reviewed in Section 2.2) are basically slight modifications of the OW I and II models. The main disadvantage of these models is that too many BS components are required to simulate ε_r and loop shapes. For instance, 10 BS components were assumed in [51] by applying the Kang model, which was too complicated from an engineering point of view. On the other hand, it is a great challenge to fit all experiments under multiple loading conditions, especially to fit those cases with the same σ_{peak} and various stress ratios ranging from <-1 to 0, as discussed in Section 4.2.2 for tests at RT and Section 5.2.2 for tests at 550 °C.

A new model is built firstly in this section based on the experimental results at RT. It is subsequently extended to fit experiments at 550 °C in Section 6.5.

One of the aims in the new modeling process is to reduce complexity: instead of four, eight, or more components of BS, the new model should include no more than four components of BS and still fulfill the 10 criteria listed in the very first place in this chapter. Among the 10 criteria, the first one, fitting on ε_r under multiple loading conditions, is the most important criterion and has the highest privilege, which means it receives the largest weight during parameter fitting.

According to the models reviewed in Section 2.2 and tested in Section 6.3, it was found that one BS component is enough to fit the shapes of stress–strain hysteresis loops. This BS component can be named BS 1, which should have a large value of plastic modulus (H_1) and a large value of modulus of dynamic recovery (C_1). The dynamic recovery rule of this BS 1 should follow the AF rule to avoid a piecewise linear shape of the hysteresis loop.

Another BS component, namely BS 2, can be assumed to control the accumulation speed of strain. This BS 2 shall have a smaller value of plastic modulus (H_2) than that of BS 1 to minimize the influence on the shape of the simulated hysteresis loop. The design of the dynamic recovery term of BS 2 is the most important task in the current modeling approach, because this term should be designed to fit ε_r under multiple loading conditions, including those with various σ_{peak}, smaller stress ratios (R < -0.8), larger stress ratios (R > -0.8), zero σ_{mean}, various $\dot{\sigma}$, and various hold times.

6.4.1 Formulation for Room Temperature

As mentioned at the beginning of this chapter, uniaxial form of model is firstly formulated, which is enough to simulate all experiments presented in Chapters 1 and 1. The equations of the constitutive model for RT tests are listed in Table 6.4.

Equation	No.
$\dot{\varepsilon} = \dot{\varepsilon}^{in} + \dot{\varepsilon}^{el}$	(6-1)
$\varepsilon^{el} = \frac{\sigma}{E}$	(6-2)
$\dot{\varepsilon}^{in} = \langle\frac{\lvert\Sigma\rvert - k}{Z}\rangle^{n}\, sgn(\Sigma)\ with\ \Sigma = \frac{\sigma}{\psi} - \Omega$	(6-3)
$\psi = \psi_1 + \psi_2, with\ \psi_1(t=0) = 0\ and\ \psi_2(t=0) = 1$	(6-5)
$\dot{\psi}_1 = -h\lvert\dot{\varepsilon}^{in}\rvert$	(6-6)
$\dot{\psi}_2 = c\left(1 - \psi_{s,\infty} - \psi_2\right)\lvert\dot{\varepsilon}^{in}\rvert$	(6-10)
$\Omega = \Omega_1 + \Omega_2$	(6-17)
$\dot{\Omega}_1 = H_1\dot{\varepsilon}^{in} - C_1\Omega_1\lvert\dot{\varepsilon}^{in}\rvert$	(6-18)
$\dot{\Omega}_2 = H_2\dot{\varepsilon}^{in} - \Omega_2\,\langle\dot{\varepsilon}^{in}\frac{\Omega_2}{r_2} + \lvert\dot{\varepsilon}^{in}\rvert\mu_2 sgn(\sigma)\rangle$	(6-19)

Table 6.4: Constitutive equations of new developed model for P91 at RT.

$E, k, Z, n, h, c, \psi_{s,\infty}, H_1, C_1, H_2, r_2$,and μ_2 are material- and temperature-dependent parameters.

The kinematic hardening (back stress) is composed of two components Ω_1 and Ω_2, as discussed at the beginning of Section 6.4. Ω_1 follows the dynamic recovery rule of Armstrong–Frederick (eq. (6-18)) and Ω_2 has a new designed dynamic recovery rule (eq. (6-19)). Instead of the parameter C_2 in eqs. (2-11), (2-14), (2-15), (2-19), (6-15), (6-16), and (6-18), the parameter r_2 is included to control the magnitude of dynamic recovery of BS 2. The parameters H_2, r_2 and μ_2 are independent from each other, in which H_2 is the plastic modulus of BS 2, and μ_2 controls the magnitude of asymmetry under tension and compression at RT.

Note that the BS 2 expresses the tendency to close the hysteresis loop, as discussed in Section 4.2.5. In Fig. 6.10, the developments of the two BS components are illustrated. It was found that the BS 2 develops around a middle stress of approximately 25 MPa, which is the mean stress of the applied stress-controlled loading. This behavior of BS 2 compensates most effects of the asymmetric loading, leaving only a tiny increment of inelastic strain in each cycle, which is the ratcheting.

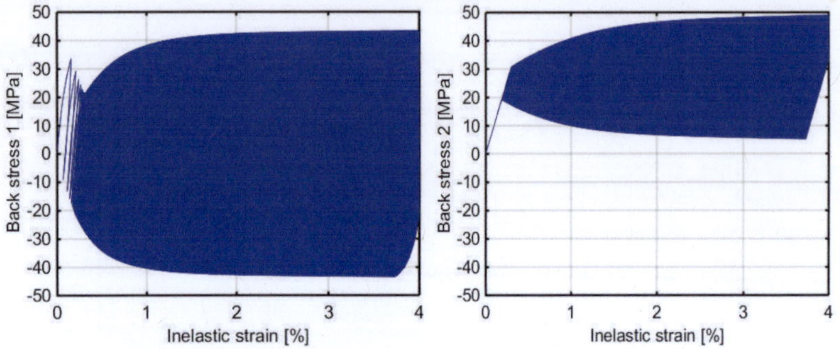

a) *BS 1 vs. inelastic strain* ***b)*** *BS 2 vs. inelastic strain.*

Fig. 6.10: Back stress components in the new developed model for the test simulated at RT with σ_{peak} = 500 MPa, σ_{mean} = 25 MPa, stress rate $\pm$50 MPa/s.

As in the Abdel–Karim model [66], various candidates for the new design of dynamic recovery rule are tested, as listed in Table 6.5.

$\Omega_2\langle\dot{\varepsilon}^{in}\rangle\langle\frac{\Omega_2}{r_2}+\mu_2\rangle$	(6-20)
$\Omega_2\lvert\dot{\varepsilon}^{in}\rvert\langle\frac{\Omega_2}{r_2}+\mu_2\rangle$	(6-21)
$\lvert\Omega_2\dot{\varepsilon}^{in}\rvert\langle\frac{\Omega_2}{r_2}+\mu_2\rangle$	(6-22)
$\langle\Omega_2\dot{\varepsilon}^{in}\rangle\langle\frac{\Omega_2}{r_2}+\mu_2\rangle$	(6-23)
$\Omega_2\left\lvert\dot{\varepsilon}^{in}\frac{\Omega_2}{r_2}+\lvert\dot{\varepsilon}^{in}\rvert\mu_2 sgn(\sigma)\right\rvert$	(6-24)

Table 6.5: Candidates of equations for dynamic recovery of back stress 2 in new developed model for P91 at RT.

Using eqs. (6-20)–(6-23) for BS 2, none of them could yield negative ratcheting for cases with σ_{max} = 500 MPa, σ_{min} = -550 MPa, and $\dot{\sigma}$ = $\pm$50 MPa/s at RT or σ_{max} = 450 MPa, σ_{min} = -500 MPa, and $\dot{\sigma}$ = $\pm$50 MPa/s at RT. Fig. 6.11 illustrates the simulation result using eq. (6-23) for BS 2. The simulated ε_r is in the opposite direction to that in the experiment. Applying eqs. (6-20)–(6-22) gives a similar model description as shown in Fig. 6.11.

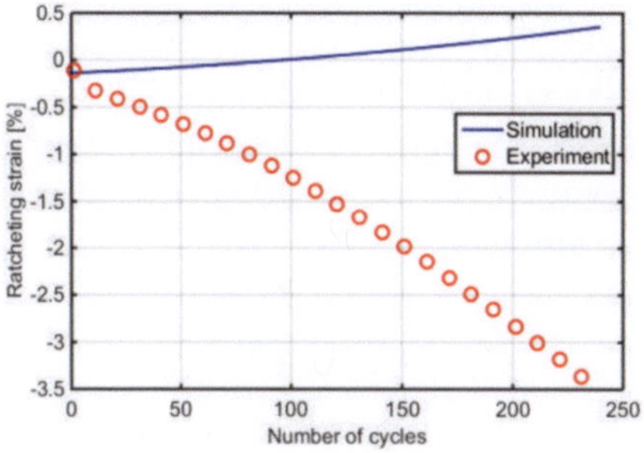

Fig. 6.11: Comparison between material response(markers) and model description(curves): Stress-controlled tests at RT with σ_{max} = 500 MPa, σ_{min} = -550 MPa, stress rate $\pm$50 MPa/s.

However, with the same parameter values, no vast difference in simulated ε_r was found between the simulation result using eq. (6-24) and the result using eq. (6-19). The difference in the model is between the Macaulay bracket ⟨x⟩ and the absolute value |x|. These two candidates were tested with difference cases, including loading with smaller stress ratios (σ_{peak} = 500 MPa, σ_{mean} = 10 and 25 MPa, and $\dot{\sigma} = \pm 50$ MPa/s), loading with larger stress ratios (σ_{peak} = 500 MPa, σ_{mean} = 100 and 250 MPa, and $\dot{\sigma} = \pm 50$ MPa/s), loading with zero σ_{mean} (σ_{peak} = 500 MPa, σ_{mean} = 0 MPa, $\dot{\sigma} = \pm 50$ MPa/s), loading with various $\dot{\sigma}$ (σ_{peak} = 500 MPa, σ_{mean} = 25 MPa, and $\dot{\sigma} = \pm 10$, 50, and 250 MPa/s), and loading with various hold times (σ_{peak} = 500 MPa, σ_{mean} = 25 MPa, $\dot{\sigma} = \pm 50$ MPa/s, one with 10 min hold time at tensile peak and another with 10 min hold time at compressive peak). Eq. (6-19) was chosen as the best among these candidates because it has better fitting to the diagram of ε_r' vs. various stress ratios, as shown in Fig. 4.13. From the mechanical point of view, application of the Macaulay bracket is more reasonable because the dynamic recovery is supposed to occur when BS and inelastic strain rate have the same direction.

The new design of dynamic recovery rule (eq. (6-19)) is similar to that in the Ohno–Abdel–Karim model [6], as shown in eqs. (2-15) and (2-16) in Chapter 1. However, the parameter μ_2 is now a controller of asymmetry under tension and compression, instead of a combination tool between the AF and OW models in [6]. With $\mu_2 > 0$, the model in cases with zero σ_{mean} yields a positive ε_r, and the model description in strain-controlled LCF tests shows higher magnitudes of compressive peak stresses than the corresponding peak tensile stresses, as already discussed in Section 4.1 and illustrated in Fig. 4.5a.

The effect of parameter μ_2 is illustrated in Fig. 6.12: In the case with σ_a = 550 MPa and zero σ_{mean}, the best simulation can be yielded with $\mu_2 = 2.4$, as shown with the blue curve in Fig. 6.12. If $\mu_2 = 0$, then the model yields no ε_r (green line in Fig. 6.12). When $\mu_2 = -2.4$, the model yields negative and symmetric ε_r (black curve in Fig. 6.12) to the simulated ε_r with $\mu_2 = 2.4$.

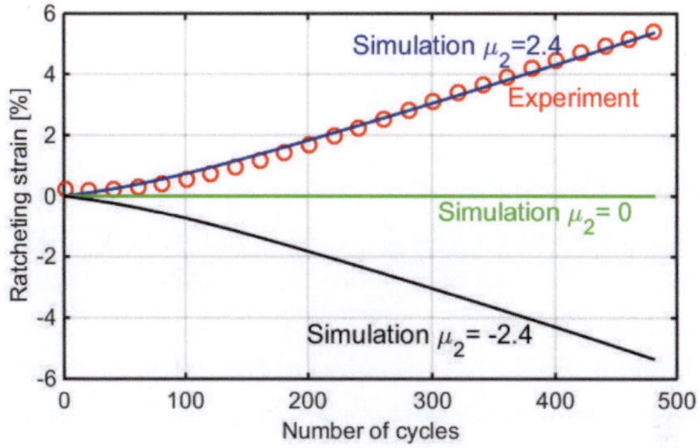

Fig. 6.12: Comparison between material response (markers) and model description (curves): Stress-controlled tests at RT performed with $\sigma_{peak} = 500$ MPa, $\sigma_{mean} = 0$ MPa, stress rate ± 50 MPa/s, with various values of μ_2.

The Yaguchi–Takahashi model [7, 26] expresses the asymmetry of material strength by introducing a term for hydrostatic pressure:

$$J_m(\boldsymbol{\sigma}) = \eta |tr(\boldsymbol{\sigma})|^l sgn(trace(\boldsymbol{\sigma})) \tag{2-21}$$

to show smaller true elastic limit under tension than that under compression. A similar fitting approach is also performed by taking $J_m(\boldsymbol{\sigma})$ into eq. (6-3), as has been done in the Yaguchi–Takahashi model (see eq. (2-22)). However, no better model description can be simulated than the application of a parameter μ_2 in the dynamic recovery term.

Hence, the positive value of μ_2 indicates that the reason for larger material strength under compression is the relatively smaller dynamic recovery. A further explanation is that the compressive hydrostatic pressure inhibits the dislocation migration and annihilation at RT, owing to second phase particles. At high temperature, for example 550 °C, dislocation migration is more thermally activated than at RT and is relatively less influenced by hydrostatic pressure, hence in the high-temperature tests reported in Chapter 1, no asymmetry was observed. However, since the asymmetry of material strength was still observed at 550 °C for mod. 9Cr–1Mo steel, as reported in [26], the suggestion above should be further verified.

The parameter values are fitted in a similar way to those in the RAFM model mentioned in Section 6.2: Young's modulus E controls the slope of the unloading line of the first cycle. The parameters k, Z, n, H_1, and C_1 control the shape of the inelastic region of the loops in strain- and stress-controlled tests. Parameter k is theoretical yield strength, parameter H_1 is inelastic modulus of BS 1, $\frac{H_1}{C_1}$ is the largest possible value of BS 1. Parameter c controls the decreasing rates of peak tensile and compressive stresses in the first stage of cyclic softening in strain-controlled tests. Parameter $\psi_{s,\infty}$ controls the magnitude of peak tensile and compressive stress at the end of the first stage of cyclic softening. Parameter h controls the slope of decrease of σ_{peak} in the saturation stage (second stage) of cyclic softening. The values of the above-mentioned parameters are fitted accordingly. The parameters H_2, r_2, and μ_2 are fitted according to $\acute{\varepsilon}_r$ under multiple cyclic loadings in stress-controlled tests presented in Section 4.2.

The trick in the parameter fitting process is to simulate $\acute{\varepsilon}_r$ under multiple loading conditions where the values of parameters k, Z, n, H_1, and C_1 still play roles. For instance, if one set of parameter values had satisfactory simulating performance for tests with high $\acute{\varepsilon}_r$, it can dramatically over- or under-estimate ε_r for tests with relatively low $\acute{\varepsilon}_r$. Different groups of values of k, Z, n, H_1, and C_1 can yield similar shaped hysteresis loops; however, these values have different performances in simulating ε_r under various stress-controlled cyclic loadings.

The same problem also exists for the parameters of softening (h, c, and $\psi_{s,\infty}$), which also affect the simulated ε_r, since cyclic softening accelerates ratcheting and this is why strain-controlled LCF tests should be performed. As illustrated in Fig. 6.13a, the simulation without softening underestimates ε_r. Further, as shown in Fig. 6.13b, the simulation result without softening shows no increase of strain range of hysteresis loop cycle by cycle, as the material does. Note that fitting on increase of strain range is also one of the criteria listed at the beginning of this chapter.

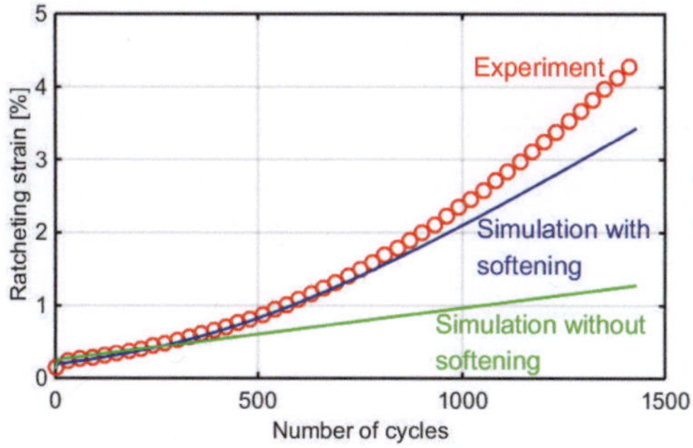

a) *ε_r vs. number of cycles.*

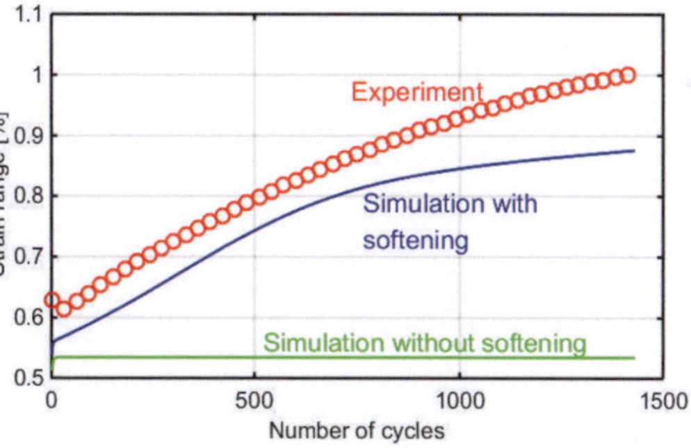

b) *Strain ranges vs. number of cycles.*

Fig. 6.13: Comparison between material response (markers) and model description (curves): Stress-controlled tests at RT performed with $\sigma_{\text{peak}} = 500$ MPa, $\sigma_{\text{mean}} = 25$ MPa, stress rate ± 50 MPa/s, with or without term of softening in simulation.

As mentioned above, the physical meaning of parameter k is the theoretical yielding strength, or the so-called true elastic limit, over which the loading can activate dislocation movement and plastic straining. However, it is impossible to measure the exact value of the true elastic limit with the current precision of the experiment equipment. The only parameter value that can be physically measured is the Young's modulus E. Therefore, except for the parameter E, all values shall be changed to fit ε_r in stress-controlled tests. The shape of loops can be compromised, if necessary, to make ε_r in the model description agree with those in material responses.

The fitted values of the 12 material parameters for P91 at RT are listed in Table 6.6. The simulation results for RT are presented in the following section.

$E\ (MPa)$	220086
$k\ (MPa)$	239.094
$Z\ (MPa\ s^{1/n})$	249.399
n	32.8131
$H_1\ (MPa)$	41551.1
C_1	787.543
$H_2\ (MPa)$	13905.9
r_2	15.3982
μ_2	0.94295
h	1.0169×10^{-3}
c	0.63815
$\psi_{s,\infty}$	0.08748

Table 6.6: Parameters of the new developed model determined for P91 at RT.

6.4.2 Simulation Results for Room Temperature

In this section, simulation results are presented for tests at RT. In each diagram, the markers indicate experimental results and the curves indicate the corresponding simulated results with the corresponding colors.

Fig. 6.14 shows the results of the strain-controlled LCF tests. The simulated shapes of the hysteresis loops approximately coincide with the experimental data points, in spite of corners near the onset points for yield (see Fig. 6.14a and b. As mentioned above, the loop shape can be compromised to fitting on ε_r in stress-controlled tests. However, the loop shapes from the simulation are still acceptable.
Fig. 6.14c shows that the cyclic softening of P91 with various strain ranges is well simulated with the new developed model. Hence, the new model satisfactorily simulates the strain-controlled tests at RT.

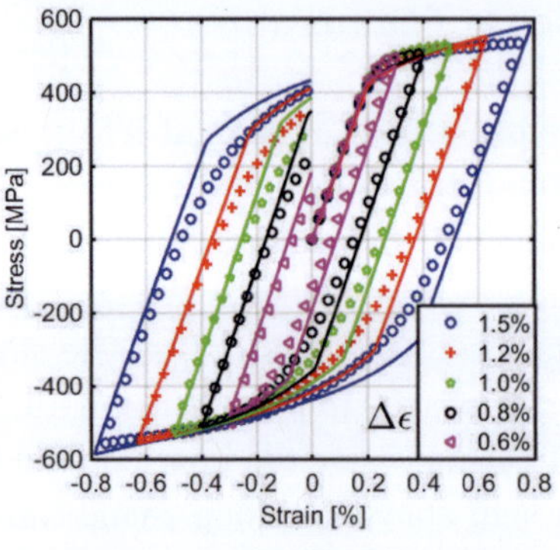

***a)** Hysteresis loops of the first cycle.*

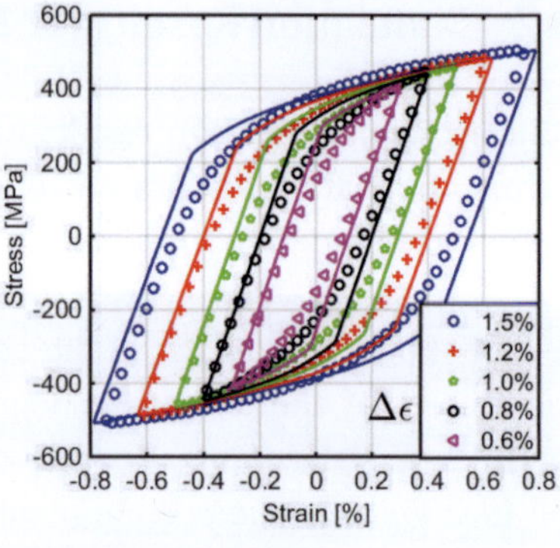

***b)** Hysteresis loops of the cycle at $N_0/2$.*

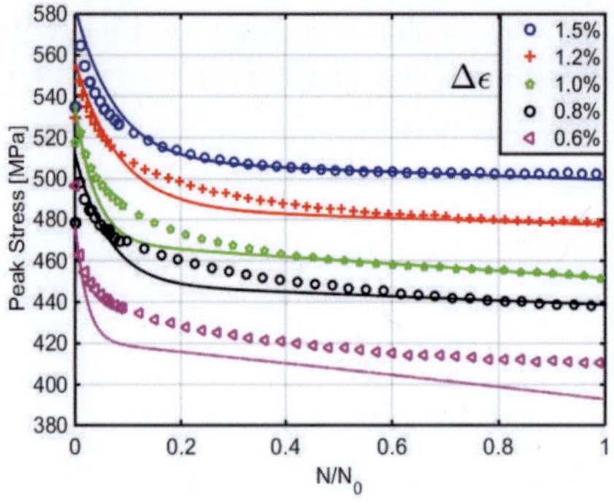

***c)** Peak tensile stresses vs. normalized number of cycles*

Fig. 6.14: Comparison between material response (markers) and model description (curves): Strain-controlled LCF tests performed with various strain ranges on P91 at RT.

Fig. 6.15 and Fig. 6.16 illustrate the simulation results for ratcheting tests performed with the same σ_{mean} and various σ_{peak}. Fig. 6.15 shows plots of ε_r vs. number of cycles for several tests. Fig. 6.16 shows the relation between $\acute{\varepsilon}_r$ and σ_{peak}. The ratcheting rates $\acute{\varepsilon}_r$ shown in Fig. 6.16, both of material responses and model description, are the average rates until the cycle in which the maximum strain reaches 4% or the cycle number achieves 2500, whichever comes first. The results show that the model is able to predict $\acute{\varepsilon}_r$ with a very large range of multiple loadings, yielding very high (> 10^{-4}) to relatively negligible (<10^{-6}) $\acute{\varepsilon}_r$ at RT.

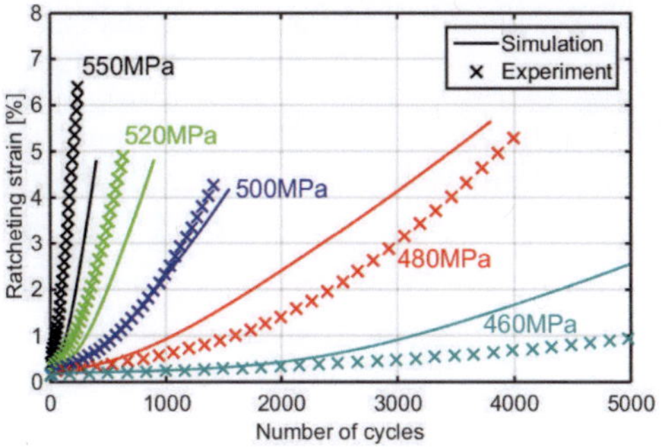

a)

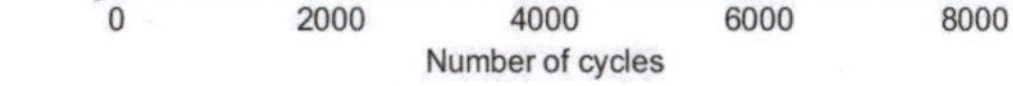

b)

Fig. 6.15: Comparison between material response (markers) and model description (curves): Ratcheting tests at RT performed with σ_{mean} = 25 MPa, stress rate ±50 MPa/s and various σ_{peak}, ε_r vs. number of cycles.

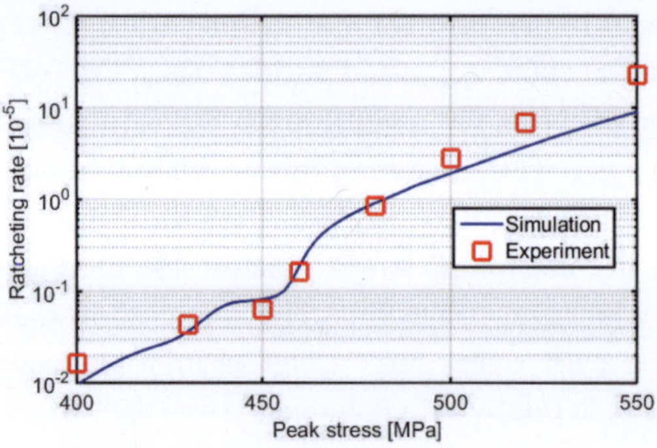

Fig. 6.16: Comparison between material response (markers) and model description (curves): Ratcheting tests at RT performed with $\sigma_{mean} = 25$ MPa, stress rate ± 50 MPa/s, various σ_{peak}, average ratcheting rates vs. σ_{peak}.

Fig. 6.17 shows the simulation results for ratcheting tests performed with the same σ_{peak} but different stress ratios. The ratcheting rates $\acute{\varepsilon}_r$ are the average rates before the 1250th cycle. As can be seen, the simulation also shows a maximum $\acute{\varepsilon}_r$ near the stress ratio of -0.9 and they show very small ε_r with stress ratios R larger than -0.6 ($-0.6 < R < 0$), similar to the material responses.

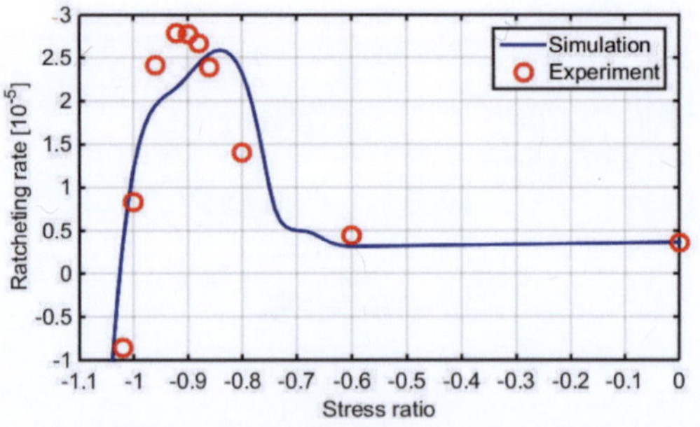

Fig. 6.17: Comparison between material response (markers) and model description (curves): Ratcheting tests at RT performed with $\sigma_{peak} = 500$ MPa, stress rate ± 50 MPa/s, various stress ratios, and average ratcheting rates vs. stress ratios.

Fig. 6.18 and Fig. 6.19 show the tests performed under symmetric stress-controlled loadings. Fig. 6.18 shows plots of ε_r vs. number of cycles with various σ_a. Fig. 6.19 shows the relation between $\acute{\varepsilon}_r$ and σ_a. The ratcheting rates $\acute{\varepsilon}_r$ are the average rates until the cycle in which the maximum strain reaches 2% or the cycle number reaches 1500, whichever comes first. The asymmetry of material strength in the model is controlled by the parameter μ_2, as discussed in Section 6.4.1. The value of μ_2 is 0.942949 (other from μ_2 = 2.4 in Fig. 6.12) to compromise between material response under different loadings. Replacing μ_2 with a function of hydrostatic pressure was also tested

$$\mu_2 = \rho|\sigma|^n \tag{6-25}$$

with ρ and n as parameters. However, no better simulation result was acquired. Therefore μ_2 was kept constant for simplicity. As shown in Fig. 6.18 and Fig. 6.19, the agreement between model description and material response is satisfactory.

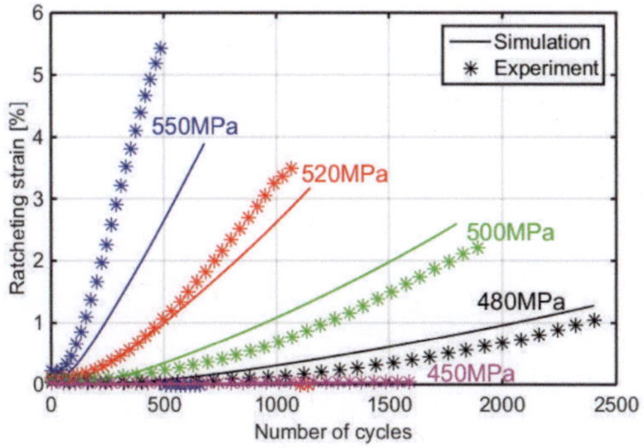

Fig. 6.18: Comparison between material response (markers) and model description (curves): Ratcheting tests at RT performed with zero σ_{mean}, stress rate ±50 MPa/s, various σ_a, and ε_r vs. number of cycles.

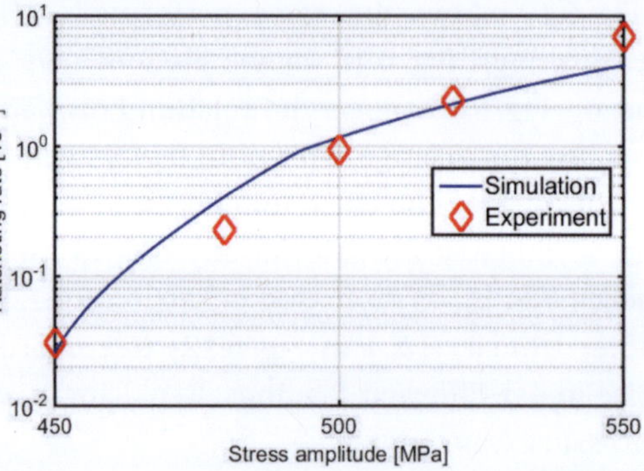

Fig. 6.19: Comparison between material response (markers) and model description (curves): Ratcheting tests at RT performed with zero σ_{mean}, stress rate ± 50 MPa/s, various σ_a, and average ratcheting rates vs. σ_a.

Fig. 6.20 illustrates the results of ratcheting tests with various $\dot{\sigma}$. The model description also shows the influence of visco-plasticity as seen in the material response. The matching between model and material responses in Fig. 6.20 is satisfactory.

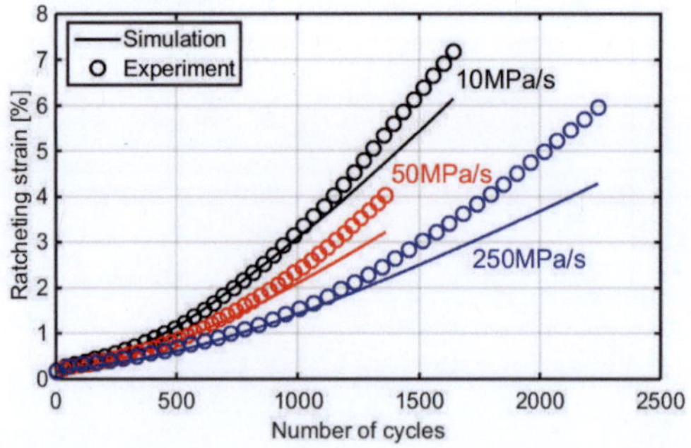

Fig. 6.20: Comparison between material response (markers) and model description (curves): Ratcheting tests at RT performed with $\sigma_{peak} = 500$ MPa, $\sigma_{mean} = 25$ MPa, various stress rates, and ε_r vs. number of cycles.

Fig. 6.21 shows those tests with 10 min hold times at peak stresses. The model perfectly simulates ε_r in the case with 10 min hold times at tensile peaks but underestimates ε_r in the case with 10 min hold times at compressive peaks. The matching, however, is still acceptable.

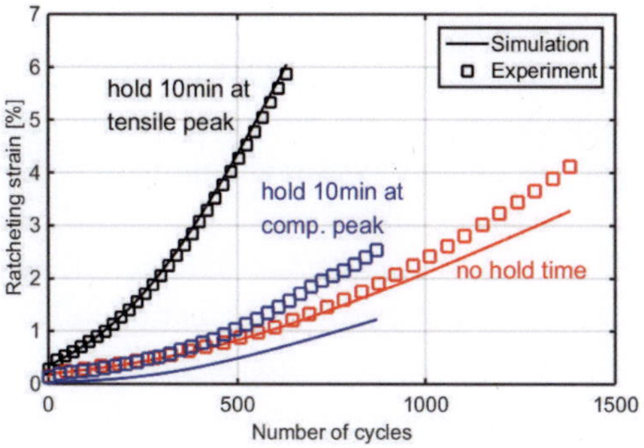

Fig. 6.21: Comparison between material response (markers) and model description (curves): Ratcheting tests at RT performed with σ_{peak} = 500 MPa, σ_{mean} = 25 MPa, various hold time types, and ε_r vs. number of cycles.

After comparing the model description and corresponding material responses in all diagrams in this section, the good simulation ability of the new developed model for P91 at RT is verified.

6.5 New Model for 550 °C

In the previous sections in this chapter, a new model was developed for P91 at RT. The simulation ability of this new developed model was proved by its model description under multiple loading conditions, as illustrated in Section 6.4.2.

Further, a model was built for P91 at 550 °C, which is based on the model for RT. The main difference for the material at RT and 550 °C is that the static recovery of kinematic hardening and isotropic softening is supposed to be thermally activated at high temperature. Hence, the corresponding terms in the constitutive equations should not be eliminated as at RT.

The modeling of BS at high temperature is basically the same as at RT, which is to keep the number of BS sub-components as two, the first one controlling the shape of simulated hysteresis loops and the second one controlling the simulated ε_r. The parameter μ_2, which expresses the asymmetry of material strength at RT, should now be zero for the high-temperature cases because, according to the experiments reported in Chapter 1, no asymmetry is observed at 550 °C. The term for dynamic recovery of the second BS component is further modified.

6.5.1 Formulation for 550 °C

The constitutive equations for P91 at 550 °C are listed in Table 6.7. Comparing to the equations for cases at RT in Table 6.4, several equations are different for the high-temperature case: in eq. (6-26), the cyclic softening controller ψ_2 includes the term expressing the static recovery of softening:

$$r_\psi |\psi_2 - \psi_r|^{m_\psi - 1} (\psi_2 - \psi_r).$$

BS 1 takes the form as in [1] (eq. (6-4)) including static recovery of kinematic hardening:

$$R_1 |\Omega_1|^{m_1 - 1} \Omega_1.$$

BS 2 (eq. (6-27)) includes the term for static recovery and the term for dynamic recovery includes the power function of the BS 2 owing to the difficulty in simulating ε_r under multiple loading conditions.

$\dot{\varepsilon} = \dot{\varepsilon}^{in} + \dot{\varepsilon}^{el}$	(6-1)
$\varepsilon^{el} = \frac{\sigma}{E}$	(6-2)
$\dot{\varepsilon}^{in} = \langle\frac{\lvert\Sigma\rvert - k}{Z}\rangle^{n}\, sgn(\Sigma)\ with\ \Sigma = \frac{\sigma}{\psi} - \Omega$	(6-3)
$\psi = \psi_1 + \psi_2, with\ \psi_1(t=0) = 0\ and\ \psi_2(t=0) = 1$	(6-5)
$\dot{\psi}_1 = -h\lvert\dot{\varepsilon}^{in}\rvert$	(6-6)
$\dot{\psi}_2 = c\left(1 - \psi_{s,\infty} - \psi_2\right)\lvert\dot{\varepsilon}^{in}\rvert - r_{\psi}\lvert\psi_2 - \psi_r\rvert^{m_{\psi}-1}(\psi_2 - \psi_r)$	(6-26)
$\Omega = \Omega_1 + \Omega_2$	(6-17)
$\dot{\Omega}_1 = H_1\dot{\varepsilon}^{in} - C_1\Omega_1\lvert\dot{\varepsilon}^{in}\rvert - R_1\lvert\Omega_1\rvert^{m_1-1}\Omega_1$	(6-4)
$\dot{\Omega}_2 = H_2\dot{\varepsilon}^{in} - \lvert\Omega_2\rvert^{n_2-1}\Omega_2\,\langle\dot{\varepsilon}^{in}\frac{\Omega_2}{r_2}\rangle - R_2\lvert\Omega_2\rvert^{m_2-1}\Omega_2$	(6-27)

Table 6.7: Constitutive equations of new developed model for P91 at 550 °C

$E, k, Z, n, h, c, r_{\psi}, \psi_r, m_{\psi}, \psi_{s,\infty}, H_1, C_1, R_1, m_1, H_2, r_2, R_2, m_2,$ and n_2 are material- and temperature-dependent parameters. Comparing the parameters for RT, there are additional ones for static recovery of cyclic softening ($r_{\psi}, \psi_r,$ and m_{ψ}) and kinematic hardening ($R_1, m_1, R_2,$ and m_2). The parameters $r_{\psi}, \psi_r,$ and m_{ψ} are fitted with experimental data from strain-controlled LCF tests with hold times, as reported in Section 5.1. The parameters $R_1, m_1, R_2,$ and m_2 are fitted with data from strain-controlled LCF tests with hold times, creep tests, and stress-controlled ratcheting tests with hold times.

Comparing the model for 550 °C and that for RT, it is obvious that the model for RT is a simplification of the model for high temperature because the static recovery of kinematic hardening and cyclic softening is eliminated for the RT case, in spite of the small difference in the term for dynamic recovery of the BS 2.

Similar to the RT case, various candidates for new design of dynamic recovery rule were tested for the 550 °C case, as listed in Table 6.8.

$\|\Omega_2\|^{n_2-1}\Omega_2\|\dot{\varepsilon}^{in}\|\frac{\Omega_2}{r_2}$	(6-28)
$\|\Omega_2\|^{n_2-1}\Omega_2\langle\dot{\varepsilon}^{in}\rangle\frac{\Omega_2}{r_2}$	(6-29)
$\|\Omega_2\|^{n_2-1}\Omega_2\left\|\dot{\varepsilon}^{in}\frac{\Omega_2}{r_2}\right\|$	(6-30)

Table 6.8: Candidates of equations for dynamic recovery of back stress 2 in new developed model for P91 at 550 °C

It was found that, by using eqs. (6-28) and (6-29), no negative ratcheting was simulated for the cases with compressive σ_{mean}. As shown in Fig. 6.22, the model description with eq. (6-28) yields positive and symmetric ε_r to the material response. The model description with eq. (6-29) is also similar.

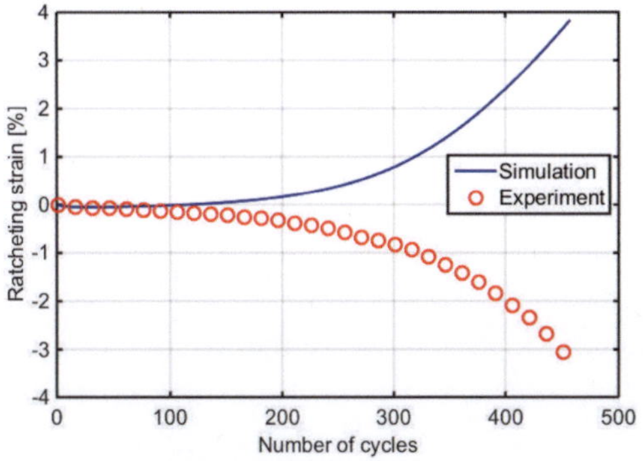

Fig. 6.22: Comparison between material response (markers) and model description (curves): Stress-controlled tests at 550°C performed with σ_{max} = 310 MPa, σ_{min} = - 325 MPa, and stress rate ±50 MPa/s.

However, no significant difference was found between the simulated results with eqs. (6-30) and (6-27). A similar situation can be recalled for the RT, in which both candidates for dynamic recovery of BS 2 (eqs. (6-19) and (6-24)) yield similar and satisfying model descriptions under multiple loading conditions, as discussed in Section 6.4.1. The difference in the models is between the Macaulay bracket $\langle x \rangle$ and the absolute value $|x|$.

Therefore, as for the RT case, the parameter values are fitted with the two candidates of equations for dynamic recovery by applying the fitting program MINUIT. Eq. (6-27) with the Macaulay bracket $\langle \dot{\varepsilon}^{in} \frac{\Omega_2}{r_2} \rangle$ is the best because it has better performance in fitting the diagram of ε_r' vs. various stress ratios, as shown in Fig. 5.14 in Chapter 1. It is also reasonable from the mechanical point of view, as discussed in Section 6.4.1 for RT.

The fitted values of the 19 material parameters for P91 at 550 °C are listed in Table 6.9. The simulation results for 550 °C are presented in the following section.

$E\ (MPa)$	173130
$k\ (MPa)$	65.2917
$Z\ (MPa\ s^{1/n})$	295.332
n	47.911
$H_1\ (MPa)$	55479.1
C_1	638.752
R_1	3.32178×10^{-3}
m_1	2.08665
$H_2\ (MPa)$	13853.8
r_2	0.21402
R_2	3.68420×10^{-6}
m_2	0.017228
n_2	0.707266
h	1.98945×10^{-2}
c	3.85381
r_ψ	8.86652×10^{-5}
ψ_r	0.850816
m_ψ	1.47506
$\psi_{s,\infty}$	0.206453

Table 6.9: Parameters of the new developed model determined for P91 at 550 °C

6.5.2 Simulation Results for 550 °C

In this section, simulation results are presented for tests at 550 °C. In each diagram, the markers indicate experimental results and the curves indicate the corresponding model description.

Fig. 6.23 shows the results in strain-controlled LCF tests at 550 °C. The simulated shapes of hysteresis loops approximately coincide with the experimental data points, in spite of not-too-sharp corners near the on-set points for yield. As in the RT cases, the loop shape can be compromised to fitting on ε_r in stress-controlled tests. However, the loop shapes in the model description are still acceptable.

Fig. 6.23c shows that the cyclic softening of P91 at high temperature with various strain ranges is well simulated with the new developed model. Hence, the new model satisfactorily simulates the strain-controlled tests at 550 °C.

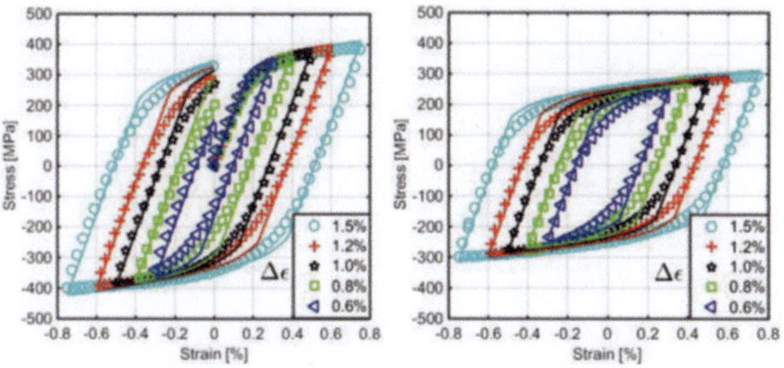

a) *Hysteresis loops of the first cycle.* ***b)*** *Hysteresis loops of the cycle at* $N_f/2$.

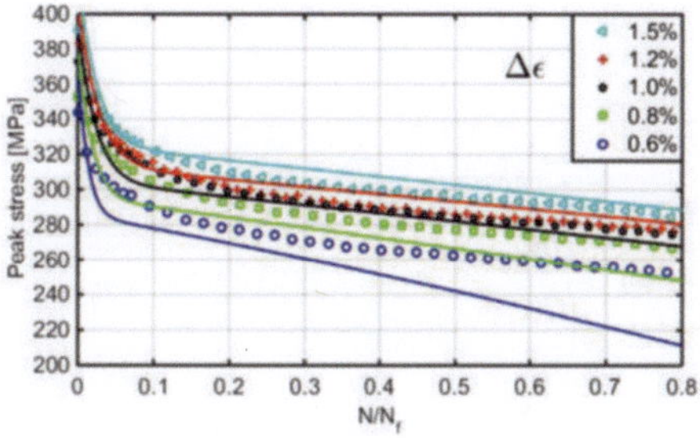

c) *Peak tensile stresses vs. normalized number of cycles*

Fig. 6.23: Comparison between material response (markers) and model description (curves): Strain-controlled LCF tests performed with various strain ranges on P91 at 550 °C.

Fig. 6.24 and Fig. 6.25 illustrate the results for ratcheting tests at 550 °C with the same σ_{mean} and different σ_{peak}. Fig. 6.24 shows plots of ε_r vs. number of cycles. Fig. 6.25 shows the relation between $\acute{\varepsilon}_r$ and σ_{peak}. The ratcheting rates $\acute{\varepsilon}_r$, both of material responses and model description, are the average rates until the cycle in which the maximum strain reaches 3%, except the marker of experiment with σ_{peak} = 275 MPa, which indicates the average $\acute{\varepsilon}_r$ when the cycle number reaches 10000, since the test was stopped at the 10000th cycle and reaches ε_r = 0.1% only. As can be seen, the agreement between model and material responses is quite good. Hence, the model is able to predict $\acute{\varepsilon}_r$ under loadings in a very large range, yielding very high to relatively negligible $\acute{\varepsilon}_r$ at 550 °C.

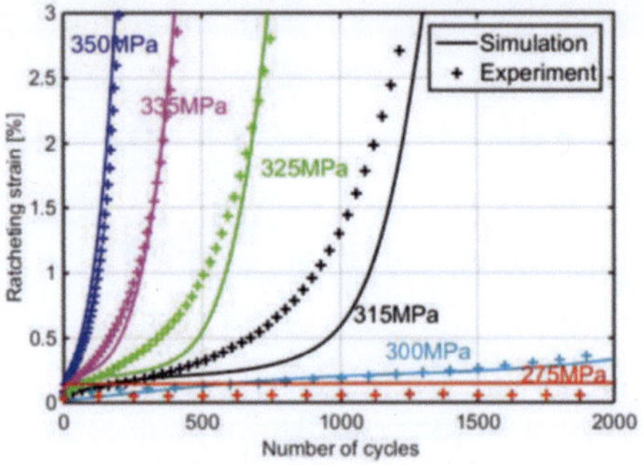

a)

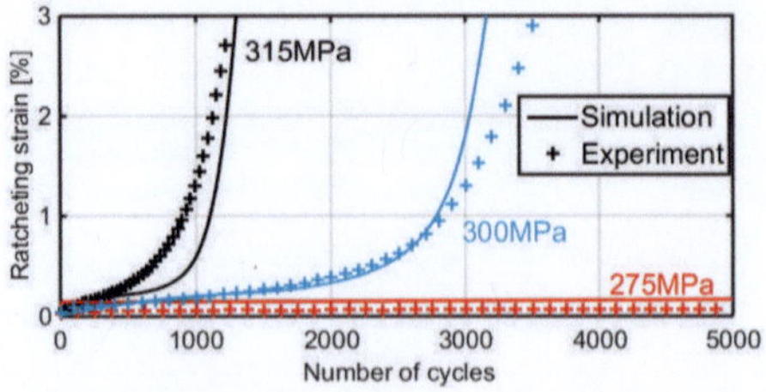

b)

Fig. 6.24: Comparison between material response (markers) and model description (curves): Ratcheting tests at 550 °C performed with σ_{mean} = 25 MPa, stress rate ±50 MPa/s, various σ_{peak}, and ε_r vs. number of cycles.

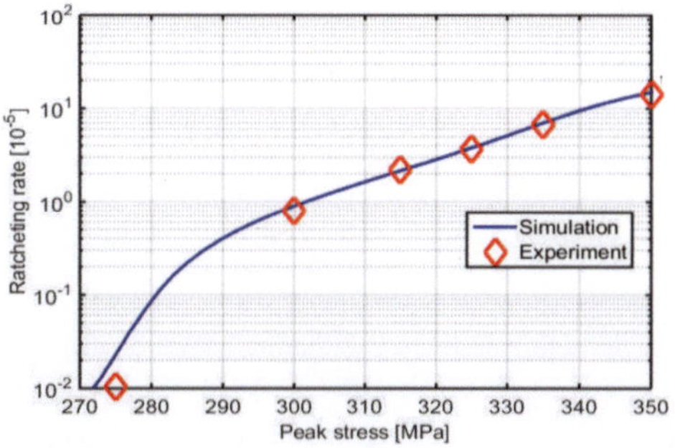

Fig. 6.25: Comparison between material response (markers) and model description (curves): Ratcheting tests at 550 °C performed with σ_{mean} = 25 MPa, stress rate ±50 MPa/s, various σ_{peak}, average ratcheting rates vs. σ_{peak}.

Fig. 6.26 shows the results of the ratcheting tests at 550 °C with the same σ_{peak} but different stress ratios. The ratcheting rates $\dot{\varepsilon}_r$ are the average rates until the cycle in which the maximum strain reaches ±3% or until the 10000th cycle, whichever comes first. As can be seen, the simulation also shows maximum $\dot{\varepsilon}_r$ near a stress ratio of -0.95 and very low $\dot{\varepsilon}_r$ with stress ratios larger than -0.6, similar to the material responses. The matching between the model and material responses is satisfactory.

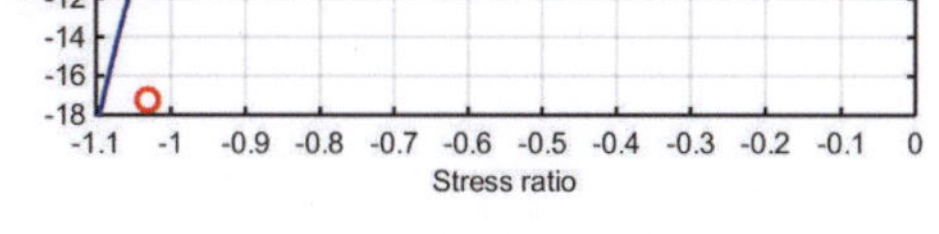

Fig. 6.26: Comparison between material response (markers) and model description (curves): Ratcheting tests at 550 °C performed with σ_{peak} = 325 MPa, stress rate ±50 MPa/s, various stress ratios, and average ratcheting rates vs. stress ratios.

Fig. 6.27 illustrates the results of ratchetings at 550 °C with various $\dot{\sigma}$. The model description also shows the influence of visco-plasticity as in the material response and simulates the material responses for the cases with stress ratios of $\pm$10 and 50 MPa very well. However, the model underestimates ε_r with a stress ratio of $\pm$250 MPa.

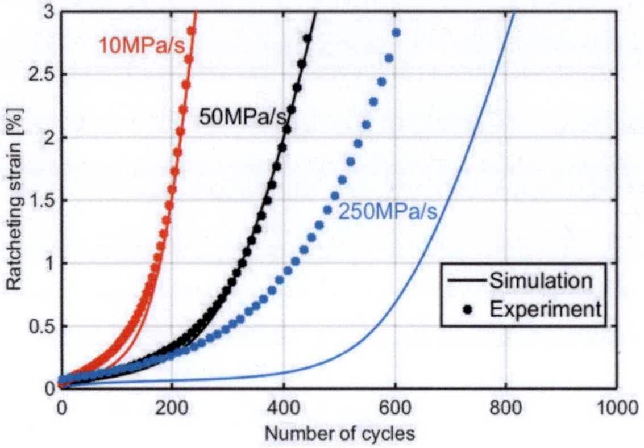

Fig. 6.27: Comparison between material response (markers) and model description (curves): Ratcheting test at 550 °C performed with σ_{peak} = 325 MPa, σ_{mean} = 7.5 MPa, various stress rates, and ε_r vs. number of cycles.

The simulated result for the case with 0.5 min hold time at 550 °C matches the material response quite well, as shown in Fig. 6.28. However, the model underestimates the accumulated strain in the case with 5 min hold time, as shown in Fig. 6.29. However, the simulated strain shown in Fig. 6.29 still increases and is only around 3–4 times slower than the experimental result. Thus, the simulation is still qualitatively correct.

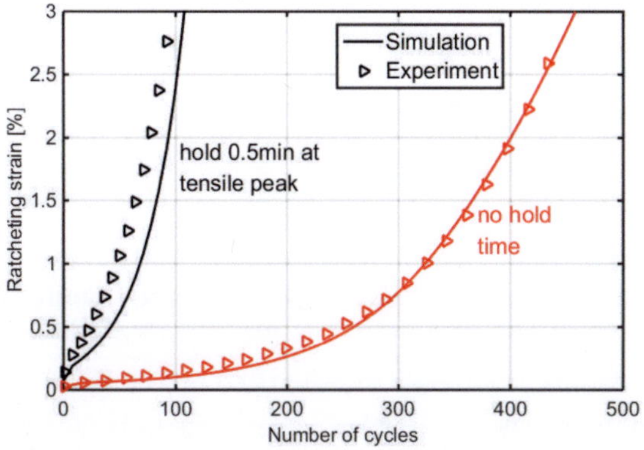

Fig. 6.28: Comparison between material response (markers) and model description (curves): Ratcheting test at 550 °C performed with σ_{peak} = 325 MPa, σ_{mean} = 7.5 MPa, stress rate ±50 MPa/s, hold 0.5 min at tensile peak and without hold time, and ε_{r} vs. number of cycles.

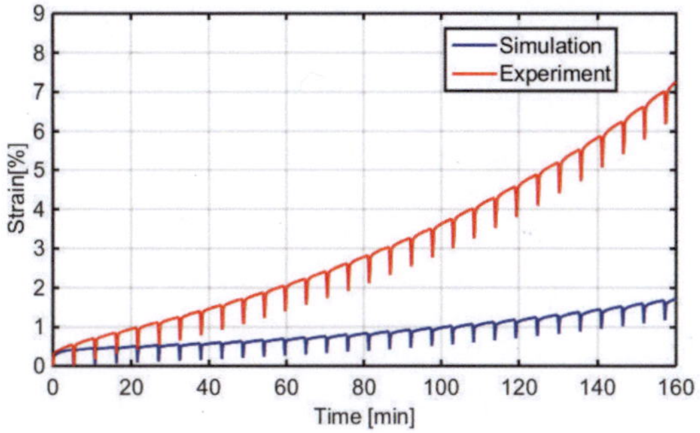

Fig. 6.29: Comparison between material response (markers) and model description (curves): Ratcheting test at 550 °C performed with σ_{peak} = 325 MPa, σ_{mean} = 7.5 MPa, stress rate ±50 MPa/s, hold 5 min at tensile peak, and strain vs. time.

After comparing the model description and corresponding material responses in all diagrams in this section, the good simulation ability of the new developed model for P91 at 550 °C is verified.

6.6 Multiaxial Formulation of the New Model

As mentioned at the beginning of this chapter, a constitutive model in uniaxial form is enough to simulate all the experiments performed in the current work, since all experiments are uniaxial.

However, in more general situations, loadings are multiaxial and the corresponding model should also be multiaxial. Although multiaxial verification experiments are unable to be carried out with the current experimental equipment, it is still worth to extend the uniaxial model to multiaxial form for the future research.

As discussed in Section 6.5.1, the constitutive model for RT is merely a simplification of the model for 550 °C. For this reason, the extension to multiaxial form can be based on the uniaxial form for 550 °C in Section 6.5.1.

The equations for the multiaxial model are listed in the Table 6.10.

$\boldsymbol{\varepsilon}=\boldsymbol{\varepsilon}^{el}+\boldsymbol{\varepsilon}^{in}$	(6-31)
$\boldsymbol{\varepsilon}^{el}=\frac{(1+v)}{E}\boldsymbol{\sigma}-\frac{v}{E}(trace(\boldsymbol{\sigma}))\mathbf{1}$	(6-32)
$\dot{\boldsymbol{\varepsilon}}^{in}=\frac{3}{2}\langle\frac{\Sigma_{eq}-K}{Z}\rangle^{n}\frac{\boldsymbol{\Sigma}}{\Sigma_{eq}}$ $with\ \boldsymbol{\Sigma}=\frac{\boldsymbol{s}}{\psi}-\boldsymbol{\Omega},\ \boldsymbol{s}=\boldsymbol{\sigma}-\frac{1}{3}trace(\boldsymbol{\sigma})\mathbf{1}\ and\ \Sigma_{eq}=\sqrt{\frac{3}{2}\boldsymbol{\Sigma}:\boldsymbol{\Sigma}}$	(6-33)
$\dot{p}=\sqrt{\frac{2}{3}\dot{\boldsymbol{\varepsilon}}^{in}:\dot{\boldsymbol{\varepsilon}}^{in}}$	(6-34)
$\psi=\psi_1+\psi_2, with\ \psi_1(t=0)=0\ and\ \psi_2(t=0)=1$	(6-5)
$\dot{\psi}_1=-h\dot{p}$	(6-6)

$\dot{\psi}_2 = c(1 - \psi_{s,\infty} - \psi_2)\dot{p} - r_\psi\lvert\psi_2 - \psi_r\rvert^{m_\psi - 1}(\psi_2 - \psi_r)$	(6-26)
$\dot{\boldsymbol{\Omega}}_1 = \frac{2}{3}H_1\dot{\boldsymbol{\varepsilon}}^{in} - C_1\boldsymbol{\Omega}_1\dot{p} - R_1 J(\boldsymbol{\Omega}_1)^{m_1 - 1}\boldsymbol{\Omega}_1$ $with\ J(\boldsymbol{\Omega}_1) = \sqrt{\frac{3}{2}\boldsymbol{\Omega}_1:\boldsymbol{\Omega}_1}$	(6-35)
$\dot{\boldsymbol{\Omega}}_2 = \frac{2}{3}H_2\dot{\boldsymbol{\varepsilon}}^{in} - J(\boldsymbol{\Omega}_2)^{n_2 - 1}\boldsymbol{\Omega}_2\langle\dot{\boldsymbol{\varepsilon}}^{in}:\frac{\boldsymbol{\Omega}_2}{r_2}$ $+\dot{p}f\rangle - R_2 J(\boldsymbol{\Omega}_2)^{m_2 - 1}\boldsymbol{\Omega}_2$ $with\ f = f(\boldsymbol{\sigma})$	(6-36)

Table 6.10: Constitutive equations of new developed model for P91 in its multiaxial formulation.

As listed in Table 6.10, since the factor indicating cyclic softening ψ is scalar, the equations for the cyclic softening (eqs. (6-5), (6-6), and (6-26)) are the same as in the uniaxial form (Table 6.7). All material parameters are the same as in the uniaxial form.

In eq. (6-36), f is a function of external loading $\boldsymbol{\sigma}$ which expresses the influence of hydrostatic pressure on the rate of dynamic recovery. One of the possible equations of f is suggested to be

$$f = \rho\lvert trace(\boldsymbol{\sigma})\rvert^n sgn(trace(\boldsymbol{\sigma})) \tag{6-37}$$

In simplification, $\rho\lvert trace(\boldsymbol{\sigma})\rvert^n$ can be replaced by a constant μ_2. Hence, by transforming eq. (6-36) into uniaxial form, $\dot{p}f$ is equivalent to $\lvert\dot{\varepsilon}^{in}\rvert\mu_2 sgn(\sigma)$ as in eq. (6-19). The exact form of f should be determined with more multiaxial tests.

Equations listed in Table 6.10 are proved to be equivalent to the uniaxial form (Table 6.4 and Table 6.7) under the uniaxial loading, according to the following calculations (eqs. (6-38)~(6-46)):

$$\boldsymbol{\sigma} = \boldsymbol{\sigma_0}\begin{pmatrix} 1 & & \\ & 0 & \\ & & 0 \end{pmatrix} \tag{6-38}$$

$$\boldsymbol{s} = \boldsymbol{\sigma_0}\begin{pmatrix} \frac{2}{3} & & \\ & -\frac{1}{3} & \\ & & -\frac{1}{3} \end{pmatrix} \tag{6-39}$$

$$\boldsymbol{\Omega} = \boldsymbol{\Omega_0}\begin{pmatrix} \frac{2}{3} & & \\ & -\frac{1}{3} & \\ & & -\frac{1}{3} \end{pmatrix} \tag{6-40}$$

$$\boldsymbol{\Sigma} = (\frac{\boldsymbol{\sigma_0}}{\boldsymbol{\psi}} - \boldsymbol{\Omega_0})\begin{pmatrix} \frac{2}{3} & & \\ & -\frac{1}{3} & \\ & & -\frac{1}{3} \end{pmatrix} \tag{6-41}$$

$$\boldsymbol{\Sigma_{eq}} = \sqrt{\frac{3}{2}\boldsymbol{\Sigma}:\boldsymbol{\Sigma}} = \left|\frac{\boldsymbol{\sigma_0}}{\boldsymbol{\psi}} - \boldsymbol{\Omega_0}\right| \tag{6-42}$$

$$\dot{\boldsymbol{\varepsilon}}^{\boldsymbol{in}} = \dot{\boldsymbol{\varepsilon}}_{\boldsymbol{0}}^{\boldsymbol{in}}\begin{pmatrix} 1 & & \\ & -\frac{1}{2} & \\ & & -\frac{1}{2} \end{pmatrix} \tag{6-43}$$

with $\dot{\boldsymbol{\varepsilon}}_{\boldsymbol{0}}^{\boldsymbol{in}} = \langle\frac{\left|\frac{\sigma_0}{\psi}-\Omega_0\right|-K}{Z}\rangle^{n}\ \boldsymbol{sgn}(\frac{\sigma_0}{\psi} - \boldsymbol{\Omega_0})$

$$\dot{\boldsymbol{p}} = \sqrt{\frac{2}{3}\dot{\varepsilon}^{in}:\dot{\varepsilon}^{in}} = \left|\dot{\boldsymbol{\varepsilon}}_{\boldsymbol{0}}^{\boldsymbol{in}}\right| \tag{6-44}$$

$$\dot{\boldsymbol{\Omega}}_1 = \dot{\boldsymbol{\Omega}}_{1,0} \begin{pmatrix} \frac{2}{3} & & \\ & -\frac{1}{3} & \\ & & -\frac{1}{3} \end{pmatrix} \tag{6-45}$$

$$\dot{\boldsymbol{\Omega}}_2 = \dot{\boldsymbol{\Omega}}_{2,0} \begin{pmatrix} \frac{2}{3} & & \\ & -\frac{1}{3} & \\ & & -\frac{1}{3} \end{pmatrix} \tag{6-46}$$

with

$$\dot{\Omega}_{2,0} = H_2 \dot{\varepsilon}_0^{in} - \left|\Omega_{2,0}\right|^{n_2-1} \Omega_{2,0} \left\langle \dot{\varepsilon}_0^{in} \frac{\Omega_{2,0}}{r_2} + \left|\dot{\varepsilon}_0^{in}\right| f \right\rangle - R_2 \left|\Omega_{2,0}\right|^{m_2-1} \Omega_{2,0}$$

7 Discussion

In the uniaxial strain-controlled and stress-controlled tests performed at RT and 550 °C, a database was built for future application of mod. 9Cr–1Mo FM steel. Under stress-controlled cyclic loading, ratcheting behavior was found on P91 steel at both temperatures. A unified visco-plastic deformation model taking into account the complex non-saturating cyclic softening of RAFM steels was further modified to adapt the ratcheting behavior of P91. The simulation ability of a new developed constitutive model was proved to be satisfactory.

However, several issues arose during this research, which should be discussed further.

One of the issues is the asymmetry of material strength under tension and compression, or namely strength-differential phenomenon. This asymmetry is observed at RT but not at 550 °C in the current work. However, [26] still reported the asymmetry of material strength at 550 °C for mod. 9Cr–1Mo steel. As mentioned in Section 2.1, several ratcheting tests reported in [26] were performed with σ_{peak} = 400 MPa, which contradicts the ultimate tensile strength of 374 MPa of T91 reported by SCK•CEN [55], although both materials mentioned in [26] and [55] are mod. 9Cr–1Mo FM steels. According to the tensile test at 550 °C in the current work, the ultimate tensile strength of P91 at 550 °C is 386.1 MPa, which is comparable to the 374 MPa for T91 reported in [55]. Hence, the tests reported in [26] should be performed with a different mod. 9Cr–1Mo FM steel than those in SCK•CEN [55] and in the current work.

An explanation for this asymmetry at RT and its disappearance at 550 °C is that the rate of dynamic recovery under compression is lower than that under tension at RT. The compressive hydrostatic pressure inhibits the dislocation migration and annihilation at RT owing to second phase particles. Hence, the hydrostatic pressure inhibits the dynamic recovery under compression. At high temperature, i.e., 550 °C, dislocation migration is more thermally activated than at RT and is relatively less influenced by hydrostatic pressure, and consequently no asymmetry is observed at high temperature. This explanation is expressed in the modeling approach with a parameter μ_2 in the term of dynamic recovery of BS 2.

Casey and Sullivan [104] and Drucker [105] suggested that the hydrostatic pressure affects the theoretical yield strength and leads to the asymmetry of material strength under tension and compression. In the modelling approaches in [104], as well as in the Yaguchi–Takahashi model [7], the hydrostatic pressure was taken to calculate the theoretical yield strength. However, the theoretical yield strength is unable to be measured without very high-precision experimental equipment. Hence, the effect of hydrostatic pressure on the theoretical yield strength has not been experimentally verified. The effect of hydrostatic pressure on the dynamic recovery can also explain the asymmetry of material strength. Accordingly, a simple introduction of a parameter μ_2 leads to satisfactory agreement between model description and material responses, as shown in Fig. 6.18 and Fig. 6.19. A similar fitting approach to the Yaguchi–Takahashi model (see eq. (2-22)) was also performed. However, the simulation results were no better than the application of parameter μ_2 in the dynamic recovery term.

On the contrary to the suggestions in [104] and [105], Jung [103] is against the suggestion that hydrostatic pressure affects dislocation movement. Jung believes that the pressure-induced increase of crystal lattice friction impeding dislocation movement completely compensates the increase of interaction between dislocations. Owing to this debate on the effect of hydrostatic pressure, further investigation on this topic is required.

According to [104] and [105], the hydrostatic pressure should lead to volumetric plastic strain of the specimens, which contradicts the prerequisite to apply the so called "true-stress-controlled ratcheting test". The true-stress controlling is based on the von Mises criterion with no volumetric plastic strain. However, after measuring the diameters of the specimens after experiments, as mentioned in Section 4.2.6, it was found that the shrinkage of the cross section area corresponds to the axial deformation, which means the volumetric plastic strain is negligible. Hence, the prerequisite to apply the "true-stress-controlled ratcheting test" was verified at RT. On the other hand, this prerequisite was only roughly satisfied at 550 °C before the total strain reaches 3% owing to necking.

The explanation for the necking at 550 °C is as follows. According to the exact measurement of the diameters along the received specimens, the diameter varies between 8.79 mm and 8.81 mm owing to deviations in production. The maximum diameter can lie anywhere within the specimen section with quasi-homogenous diameter. However, the necking always occurs exactly at the middle of the specimens, as shown in Fig. 5.22, which indicates that the middle of the specimen is always softer than the other sections. According to temperature measurement along the specimen, when the middle section is 550 °C, the two ends of the quasi-homogenous section are around 545 °C, which is caused by the temperature distribution in the whole oven. Hence, the temperature difference can explain why the necking always occurs at exactly the middle of the specimen. Further, the data acquired in the experiments at high temperature should be an average value within the gage length of the extensometer.

The definition of lifetime of specimens should be reconsidered. The lifetime of a structural component is literally a period of usage or number of usage after which the component is no longer safe for further application and must be replaced. However, the criteria to define such period or number of usage are not universal. In the strain-controlled LCF tests at 550 °C, the definition of lifetime follows the way reported in [101], which is the cycle during which σ_{peak} has decreased by 10% from that predicted by extrapolation of the saturation curve (stage 2), in which an obvious macro crack propagation occurs.

However, in spite of measurement error, as long as σ_{peak} clearly deviates from the extrapolation of the saturation curve, that is, decreases by only 1% from the extrapolation line, it is clear that unstable macro crack propagation has already begun. Hence it is supposed the 10% decrease from the extrapolation line reported in [101] is only to eliminate the measurement error to be certain that the macro cracking has already started. In the strain-controlled LCF tests of the current work, the curves in σ_{peak}-N diagrams (as shown in Fig. 4.3 and Fig. 5.4) are generally smoother than those reported in [101] (as shown in Fig. 2.12.). Hence, in the current work, it is also possible to define the lifetime N_f in the LCF tests at 550 °C by the cycle with, for example, 5% deviation from the extrapolation line.

In the LCF tests at RT, a clear definition of lifetime is not possible since the macro crack does not always appear within the gage length of the extensometer. For this reason, data are considered to be reliable until a roughly defined cycle N_0, which lies within the stage where the peak stress is linearly decreasing (saturation stage of cyclic softening) and near the cycle where the deviation of peak stress from the linear decrease appears owing to the macro crack. Further, the lifetime N_f is roughly determined as the cycle number N_0. Although it is roughly defined, there is no vast difference between this cycle number N_0 to a clearly defined N_f with, for example, a 10% decrease from the extrapolation line.

In the stress-controlled tests, on the other hand, the failure of the specimen is generally not owing to macro cracking, but rather unacceptable deformation of, for example, 5% owing to ratcheting, as in most of the ratcheting tests reported in Chapters 1 and 1. The criterion for an "unacceptable deformation" can only be clearly determined in practical application. However, in several stress-controlled tests in the current work, for example, the test performed with σ_a = 520 MPa at RT and the four tests at 550 °C (σ_a = 350, 340, 325, and 315 MPa) under symmetric stress-controlled loading, the macro crack still appears before 5% deformation owing to ratcheting.

Further, owing to the fact that necking occurs on most specimens tested at 550 °C, as shown in Fig. 5.22, the failure of the specimens is that they are no longer safe after obvious shrinkage from the cross section as received, for example, 5%. As mentioned above, the reason for the necking is suggested to be distribution of temperature along the specimen. Hence, the definition of lifetime in these cases is even more complicated because necking will lead to either unacceptable deformation or fracture very fast.

It is recalled that in the tests with hold times at RT, as reported in Section 4.2.5, there is inelastic deformation during hold times. In the test performed with static stress σ = 500 MPa, there is an inelastic strain increment of 0.24% in the first 5 hours after the stress stops at 500 MPa and only 0.02% inelastic strain increment during the following 19 hours. The inelastic strain with static stress at RT is similar to creep, although creep is generally understood to obviously occur when the temperature is higher than approximately 0.3 of the melting temperature. However, there is no exactly defined temperature above which creep will happen. In the test with static stress σ = 500 MPa, it is believed that it is a creep behavior with only a primary stage and no secondary stage.

One more phenomenon in the experiment is worth discussing: Comparing the tests reported in Section 4.2.4 and 5.2.4, it was found that the ratcheting test performed with stress rate $\pm$10 MPa has an 18.3% higher $\acute{\varepsilon}_r$ than that with stress rate $\pm$50 MPa at RT, but is twice as high at 550 °C, which indicates lower viscosity at high temperature. However, the test performed with $\pm$250 MPa/s has a 22.3% lower $\acute{\varepsilon}_r$ than the test performed with $\pm$50 MPa/s at RT while it is only 26.5% lower at 550 °C. It was found the percentage changes of 22.3% and 26.5% were comparable. This phenomenon implies that, at high loading rate, the viscosity is relatively less sensitive to the change of temperature.

There are also several issues in question in the modeling approach of the current work. The current modeling approach is similar to that in Koo and Lee [24], which was also for mod. 9Cr–1Mo steel. The number of BS components was reduced to three in their report [24]. It was found that the term of dynamic recovery of the third BS component could control the rate of ratcheting. However, the ratcheting tests reported in [24] were too few to cover a large range of loading conditions and all ratcheting tests were stopped before the 100th cycle. Further, it is questionable that, in the ratcheting tests reported in [24], $\acute{\varepsilon}_r$ went progressively slower after the so-called "progressive deformation instability", as shown in Fig. 2.3 and Fig. 2.4, which indicates this "instability" happened only once and disappeared afterwards. Such "instability" was only intuitively suggested to be the consequence of cyclic softening, however without explicit explanation or any approach to simulate this "instability" with the proposed model. In the tests at 550 °C, which were reported in Chapter 1, such "instability" was never observed. The simulation results of ratcheting were not illustrated in [24] with their proposed model. Besides, since the asymmetry of material strength at 600 °C was not observed, the proposed model in [24] did not include a term to express such asymmetry. For these reasons, it was necessary to develop a new constitutive model, as has been reported in Chapter 1, in which a model was developed with only two BS components, one controlling the shapes of hysteresis loops and the other controlling the ratcheting rates.

When comparing the fitted parameter values listed in Table 6.3 and those in Table 6.6, it can be seen that the values of parameters k, Z, n, H_1, C_1, h, c, and $\psi_{s,\infty}$ are different in the two tables. As mentioned in Section 6.4.1, different groups of these parameter values can yield similar shaped hysteresis loops in simulation of either strain- or stress-controlled tests. However, these groups of parameter values have different performances in simulating ε_r under multiple stress-controlled cyclic loadings. Hence, although parameters have their own physical meaning, as mentioned in Section 6.4.1, it is only possible to fit the parameters until an acceptable simulation is acquired, which means that a direct determination of these parameter values, except for Young's modulus, is still not possible.

In the choice of a better constitutive equation of BS 2 for both RT and 550 °C cases, it comes to the choice between the Macaulay bracket (eqs. (6-19) and (6-27)) and the absolute value (eqs. (6-24) and (6-30)). As mentioned in Section 6.4.1, the Macaulay bracket is more reasonable based on the hypothesis that the dynamic recovery is supposed to occur only when BS and inelastic strain rate having the same direction. This hypothesis follows the modeling approach of the OW I model, as shown in eqs. (2-11), (2-12), and (2-13). However, according to the AF model, dynamic recovery would also occur when BS and inelastic strain rate have opposite directions, as shown in eq. (2-6) and eq. (2-7). The physical background was not discussed in Ohno and Wang's report [3, 4], or in Armstrong and Frederick's report [62]. Ohno and Abdel–Karim [6] simply combined both ideas into eq. (2-15) without further discussion from the physical point of view.

The section in which the BS and inelastic strain rate having opposite signs is illustrated in red in the stress–strain hysteresis loop shown in Fig. 7.1. In spite of the difference from the physical point of view, the simulated results are similar when using the Macaulay bracket and the absolute value. Fig. 7.2 illustrates the comparison of increasing rate of BS 2 between application of the Macaulay bracket (blue curve) and the absolute value (red curve). As can be seen, the blue and red curves coincide with each other, except for two short sections, and the deviation is relative small. Hence, the simulated results are similar.

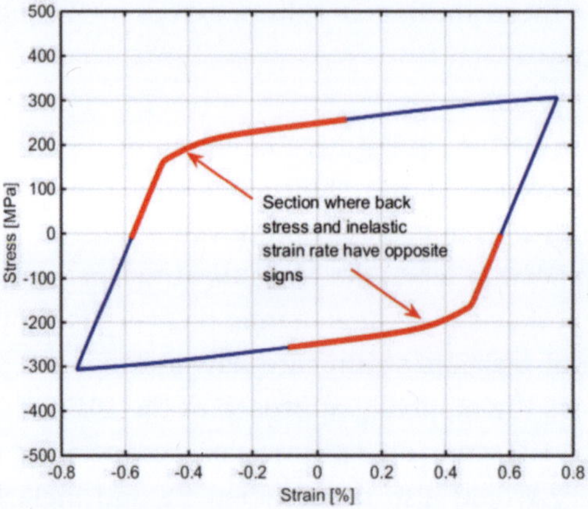

Fig. 7.1: Simulated stress–strain hysteresis loop of 200th cycle in the strain-controlled LCF test performed with $\Delta\varepsilon$ = 1.5% at 550 °C

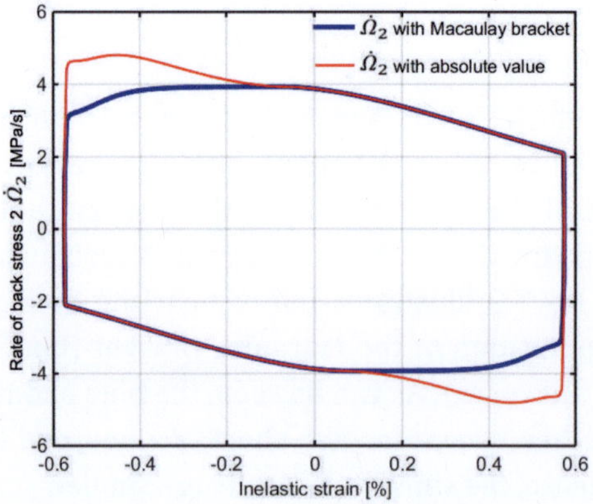

Fig. 7.2: Comparison between using Macaulay bracket and absolute value, simulated result: 200th cycle of the strain-controlled LCF test performed with $\Delta\varepsilon$ = 1.5% at 550 °C, increasing rate of BS 2 vs. inelastic strain.

The last, but not least, issue is that the new proposed model in the current work does not have satisfactory performance for simulating the strain increment in the ratcheting test at 550 °C with long hold times (i.e., 5 min) at peak stress. The reason is that in parameter fitting, the criterion "fitting on ratcheting strains in all experiments" has the largest weight while most experiments at 550 °C are performed without or with a short hold time (0.5 min). However, in the experiment illustrated in Fig. 6.29, the total strain is mostly composed of accumulated creep strains during hold times, which is a different mechanism of accumulation of inelastic strain, as discussed in Section 0. Hence, the criterion "fitting on accumulation of strain in creep test" compromises for the "fitting on ratcheting strain". Further research is required to investigate the inelastic strain increment of P91 during creep or hold time at 550 °C.

8 Summary and Outlook

In the uniaxial strain-controlled and stress-controlled tests performed at RT and 550 °C, a database was built for future application of 9Cr–1Mo FM steel at RT and 550 °C. The material under research shows obvious cyclic softening in strain-controlled LCF tests at both temperatures. (see Fig. 4.2c and Fig. 5.3c).

At RT, the material shows asymmetry of material strength under tension and compression, which is supported by three facts:
a) Larger magnitude of peak compressive stress than corresponding peak tensile stress in strain-controlled LCF tests (see Fig. 4.4b);
b) Lower magnitude of ratcheting rate with compressive mean stress in stress-controlled ratcheting tests (see Fig. 4.10c);
c) Positive ratcheting under symmetric stress-controlled loadings (see Fig. 4.15).

At 550 °C, the material shows symmetry of material strength under tension and compression, which is supported by three facts:
a) Identical magnitude of peak compressive and tensile stress in strain-controlled LCF tests (see Fig. 5.5);
b) Identical magnitudes of ratcheting rates with symmetric compressive and tensile mean stress (see Fig. 5.11);
c) Negligible ratcheting under symmetric stress-controlled loadings (see Fig. 5.15).

In spite of the difference in the symmetry of material strength at RT and 550 °C, the effects of various influencing factors to $\acute{\varepsilon}_r$ are similar:

1. With the same σ_{mean}, larger σ_{peak} leads to higher $\acute{\varepsilon}_r$ (see Fig. 4.7 and Fig. 5.8).

2. With the same σ_{peak}, highest $\acute{\varepsilon}_r$ is reached with stress ratio of around -0.9--0.95 (see Fig. 4.13 and Fig. 5.14).
3. Strain range of hysteresis loop increases with cyclic number (see Fig. 4.8 and Fig. 5.9).
4. Higher $\dot{\sigma}$ leads to lower $\acute{\varepsilon}_r$ and vice versa (see Fig. 4.18 and Fig. 5.17).

A constitutive model has been proposed. The model is originated from the Chaboche Model and adopts isotropic softening factor from the RAFM model. To simulate the ratcheting behavior of the material, the kinematic hardening/back stress (BS) is divided into two sub-components, one controls the shapes of simulated stress–strain hysteresis loops, and the other controls the ratcheting rate. The asymmetry of material strength at RT is expressed with a parameter μ_2. This model has been proved to have good simulation ability for material response under both strain- and stress-controlled loading conditions. The simulated results agree with the corresponding material response within a large range of multiple loading.

Since all tests in the current work were performed only under uniaxial loading, the proposed model is in uniaxial form, which is enough to generate simulated results to compare with performed experiments. Owing to the fact that the multiaxial loadings are more general in the real working condition of the structural components, further extension to multiaxial form of the proposed uniaxial model was also proposed. In the further research on P91 steel, multiaxial tests shall be performed to verify the model in multiaxial form.

In the current work, only isothermal mechanical behavior at two temperatures (RT and 550 °C) was investigated. Tests shall be performed at more temperatures in future research to check whether the asymmetry of material strength under tension and compression at RT still exists. Microstructural investigation is required to discover the cause of such asymmetry at RT. Further, research on thermal mechanical behavior shall be carried out at conditions closer to the real working conditions of the structural components

Since the failure mode of the specimens in the ratcheting test is supposed to be unacceptable ratcheting strain, no damage factor is included in the proposed constitutive model. In future improvement of the model, more tests should be performed that end with fracture to investigate the damage of the material.

9 Reference

[1] J. Aktaa and R. Schmitt, "High temperature deformation and damage behavior of RAFM steels under low cycle fatigue loading: Experiments and modeling", *Fusion Engineering and Design,* vol. 81, pp. 2221-2231, 2006.

[2] H. Hübel, "Basic conditions for material and structural ratcheting", *Nuclear Engineering and Design,* vol. 162, pp. 55-65, 1996.

[3] N. Ohno and J. D. Wang, "Kinematic Hardening Rules with Critical State of Dynamic Recovery, Part I: Formulation and basic features for ratchetting behavior", *International Journal of Plasticity,* vol. 9, pp. 375-390, 1993.

[4] N. Ohno and J. D. Wang, "Kinematic Hardening Rules with Critical State of Dynamic Recovery, Part II: Application to experiments of ratcheting behavior", *International Journal of Plasticity,* vol. 9, pp. 391-403, 1993.

[5] J. L. Chaboche and G. Cailletaud, "On the calculation of structures in cyclic plasticity or viscoplasticity", *Computers & Structures,* vol. 23, pp. 23-31, 1986.

[6] M. Abdel-Karim and N. Ohno, "Kinematic hardening model suitable for ratchetting with steady-state", *International Journal of Plasticity,* vol. 16, pp. 225-240, 2000.

[7] M. Yaguchi and Y. Takahashi, "Ratchetting of viscoplastic material with cyclic softening, part 2: application of constitutive models", *International Journal of Plasticity,* vol. 21, pp. 835-860, 2005.

[8] L. F. Coffin, "The Influence of Mean Stress on the Mechanical Hysteresis Loop Shift of 1100 Aluminum", *Journal of Fluids Engineering,* vol. 86(4), pp. 673-680, 1964.

[9] G. Facheris and K. G. F. Janssens, "An internal variable dependent constitutive cyclic plastic material description including

ratcheting calibrated for AISI 316L", *Computational Materials Science,* vol. 87, pp. 160-171, 2014.

[10] G. Facheris and K. G. F. Janssens, "Cyclic mechanical behavior of 316L: Uniaxial LCF and strain-controlled ratcheting tests", *Nuclear Engineering and Design,* vol. 257, pp. 100-108, 2013.

[11] Y. Dong, G. Kang, Y. Liu, H. Wang, and X. Cheng, "Dislocation evolution in 316L stainless steel during multiaxial ratchetting deformation", *Materials Characterization,* vol. 65, pp. 62-72, 2012.

[12] A. Sarkar, A. Nagesha, P. Parameswaran, R. Sandhya, and M. D. Mathew, "Influence of dynamic strain aging on the deformation behavior during ratcheting of a 316LN stainless steel", *Materials Science and Engineering: A,* vol. 564, pp. 359-368, 2013.

[13] S. Krishna, T. Hassan, I. Ben Naceur, K. Saï, and G. Cailletaud, "Macro versus micro-scale constitutive models in simulating proportional and nonproportional cyclic and ratcheting responses of stainless steel 304", *International Journal of Plasticity,* vol. 25, pp. 1910-1949, 2009.

[14] G. Kang, Q. Gao, and X. Yang, "Uniaxial and non-proportionally multiaxial ratcheting of SS304 stainless steel at room temperature: experiments and simulations", *International Journal of Non-Linear Mechanics,* vol. 39, pp. 843-857, 2004.

[15] G. Kang, Q. Gao, L. Cai, and Y. Sun, "Experimental study on uniaxial and nonproportionally multiaxial ratcheting of SS304 stainless steel at room and high temperatures", *Nuclear Engineering and Design,* vol. 216, pp. 13-26, 2002.

[16] Q. Gao, G. Z. Kang, and X. J. Yang, "Uniaxial ratcheting of SS304 stainless steel at high temperatures: visco-plastic constitutive model", *Theoretical and Applied Fracture Mechanics,* vol. 40, pp. 105-111, 2003.

[17] G. Kang, Y. Li, and Q. Gao, "Non-proportionally multiaxial ratcheting of cyclic hardening materials at elevated temperatures: Experiments and simulations", *Mechanics of Materials,* vol. 37, pp. 1101-1118, 2005.

[18] Y. Liu, G. Kang, and Q. Gao, "Stress-based fatigue failure models for uniaxial ratchetting–fatigue interaction", *International Journal of Fatigue,* vol. 30, pp. 1065-1073, 2008.

[19] G. Kang, Q. Gao, and X. Yang, "A visco–plastic constitutive model incorporated with cyclic hardening for uniaxial/multiaxial ratcheting of SS304 stainless steel at room temperature", *Mechanics of Materials,* vol. 34, pp. 521-531, 2002.

[20] G. Kang and Q. Kan, "Constitutive modeling for uniaxial time-dependent ratcheting of SS304 stainless steel", *Mechanics of Materials,* vol. 39, pp. 488-499, 2007.

[21] M. Shariati and H. Hatami, "Experimental study of SS304L cylindrical shell with/without cutout under cyclic axial loading", *Theoretical and Applied Fracture Mechanics,* vol. 58, pp. 35-43, 2012.

[22] C. K. Mukhopadhyay, T. Jayakumar, T. K. Haneef, S. Suresh Kumar, B. P. C. Rao, S. Goyal, *et al.*, "Use of acoustic emission and ultrasonic techniques for monitoring crack initiation/growth during ratcheting studies on 304LN stainless steel straight pipe", *International Journal of Pressure Vessels and Piping,* vol. 116, pp. 27-36, 2014.

[23] *304, 304L, 304H, 304N, 304HN, 304LN 18-8 Austenitic Stainless Steel.* Available: http://amsresources.com/304-austenitic-stainless-steel/

[24] G.-H. Koo and J.-H. Lee, "Investigation of ratcheting characteristics of modified 9Cr–1Mo steel by using the Chaboche constitutive model", *International Journal of Pressure Vessels and Piping,* vol. 84, pp. 284-292, 2007.

[25] X.-T. Zheng, F.-Z. Xuan, and P. Zhao, "Ratcheting-creep interaction of advanced 9–12% chromium ferrite steel with anelastic effect", *International Journal of Fatigue,* vol. 33, pp. 1286-1291, 2011.

[26] M. Yaguchi, "Ratchetting of viscoplastic material with cyclic softening, part 1: experiments on modified 9Cr-1Mo steel", *International Journal of Plasticity,* vol. 21, pp. 43-65, 2005.

[27] P. Zhao and F.-Z. Xuan, "Ratchetting behavior of advanced 9–12% chromium ferrite steel under creep-fatigue loadings", *Mechanics of Materials,* vol. 43, pp. 299-312, 2011.

[28] M. Ando, N. Isobe, K. Kikuchi, and Y. Enuma, "Effect of ratchet strain on fatigue and creep–fatigue strength of Mod.9Cr–1Mo steel", *Nuclear Engineering and Design,* vol. 247, pp. 66-75, 2012.

[29] Y. Li, X. Pan, and G. Wang, "Low cycle fatigue and ratcheting properties of steel 40Cr under stress controlled tests", *International Journal of Fatigue,* vol. 55, pp. 74-80, 2013.

[30] G. Kang, Y. Liu, J. Ding, and Q. Gao, "Uniaxial ratcheting and fatigue failure of tempered 42CrMo steel: Damage evolution and damage-coupled visco-plastic constitutive model", *International Journal of Plasticity,* vol. 25, pp. 838-860, 2009.

[31] G. Kang, Y. Liu, and J. Ding, "Multiaxial ratchetting–fatigue interactions of annealed and tempered 42CrMo steels: Experimental observations", *International Journal of Fatigue,* vol. 30, pp. 2104-2118, 2008.

[32] C. Li, G. Chen, X. Chen, and W. Zhang, "Ratcheting strain and simulation of 16MnR steel under uniaxial cyclic loading", *Computational Materials Science,* vol. 57, pp. 43-47, 2012.

[33] M. Zeinoddini and M. Peykanu, "Strain ratcheting of steel tubulars with a rectangular defect under axial cycling: A numerical modeling", *Journal of Constructional Steel Research,* vol. 67, pp. 1872-1883, 2011.

[34] G. Kang, Y. Dong, Y. Liu, H. Wang, and X. Cheng, "Uniaxial ratchetting of 20 carbon steel: Macroscopic and microscopic experimental observations", *Materials Science and Engineering: A,* vol. 528, pp. 5610-5620, 2011.

[35] Y. Dong, G. Kang, Y. Liu, and H. Jiang, "Multiaxial ratcheting of 20 carbon steel: Macroscopic experiments and microscopic observations", *Materials Characterization,* vol. 83, pp. 1-12, 2013.

[36] X. Yang, "Low cycle fatigue and cyclic stress ratcheting failure behavior of carbon steel 45 under uniaxial cyclic loading", *International Journal of Fatigue,* vol. 27, pp. 1124-1132, 2005.

[37] N. Khutia, P. P. Dey, S. K. Paul, and S. Tarafder, "Development of non Masing characteristic model for LCF and ratcheting fatigue simulation of SA333 C–Mn steel", *Mechanics of Materials,* vol. 65, pp. 88-102, 2013.

[38] S. K. Paul, "Effect of anisotropy on ratcheting: An experimental investigation on IFHS steel sheet", *Materials Science and Engineering: A,* vol. 538, pp. 349-355, 2012.

[39] M. Wen, H. Li, D. Yu, G. Chen, and X. Chen, "Uniaxial ratcheting behavior of Zircaloy-4 tubes at room temperature", *Materials & Design,* vol. 46, pp. 426-434, 2013.

[40] J. Peng, C.-Y. Zhou, Q. Dai, X.-H. He, and X. Yu, "Fatigue and ratcheting behaviors of CP-Ti at room temperature", *Materials Science and Engineering: A,* vol. 590, pp. 329-337, 2014.

[41] L.-x. Cai, Q.-y. Niu, S.-y. Qiu, and Y.-j. Liu, "Ratcheting Behavior of T225NG Titanium Alloy under Uniaxial Cyclic Stressing: Experiments and Modeling", *Chinese Journal of Aeronautics,* vol. 18, pp. 31-39, 2005.

[42] Q. Kan and G. Kang, "Constitutive model for uniaxial transformation ratchetting of super-elastic NiTi shape memory alloy at room temperature", *International Journal of Plasticity,* vol. 26, pp. 441-465, 2010.

[43] G. Kang, "Advances in transformation ratcheting and ratcheting-fatigue interaction of NiTi shape memory alloy", *Acta Mechanica Solida Sinica,* vol. 26, pp. 221-236, 2013.

[44] M. Kotoul, "Constitutive modeling of ratcheting of metal particulate-reinforced ceramic matrix composites", *materials Science and Engineering: A,* vol. 319-321, pp. 657-661, 2001.

[45] Z. Zhang and X. Chen, "Multiaxial ratcheting behavior of PTFE at room temperature", *Polymer Testing,* vol. 28, pp. 288-295, 2009.

[46] A. D. Drozdov and J. d. Christiansen, "Cyclic viscoelastoplasticity of polypropylene/nanoclay hybrids", *Computational Materials Science,* vol. 53, pp. 396-408, 2012.

[47] M. Shariati, H. Hatami, H. Yarahmadi, and H. R. Eipakchi, "An experimental study on the ratcheting and fatigue behavior of polyacetal under uniaxial cyclic loading", *Materials & Design,* vol. 34, pp. 302-312, 2012.

[48] G. Tao and Z. Xia, "Ratcheting behavior of an epoxy polymer and its effect on fatigue life", *Polymer Testing,* vol. 26, pp. 451-460, 2007.

[49] X. Chen and R. Jiao, "Modified kinematic hardening rule for multiaxial ratcheting prediction", *International Journal of Plasticity,* vol. 20, pp. 871-898, 2004.

[50] T. Hassan, L. Taleb, and S. Krishna, "Influence of non-proportional loading on ratcheting responses and simulations by two recent cyclic plasticity models", *International Journal of Plasticity,* vol. 24, pp. 1863-1889, 2008.

[51] G. Kang, "A visco-plastic constitutive model for ratcheting of cyclically stable materials and its finite element implementation", *Mechanics of Materials,* vol. 36, pp. 299-312, 2004.

[52] S. S. Manson, "Behavior of Materials under Conditions of Thermal Stress," Lewis Flight Propulsion Laboratory1952.

[53] R. W. Smith, M. H. Hirschberg, and S. S. Manson, "Fatigue behavior of materials under strain cycling in low and intermediate life range," NASA1978.

[54] J. L.F. Coffin, "The Deformation and Fracture of a Ductile Metal Under Superimposed Cyclic and Monotonic Strain," in *Achievement of High Fatigue Resistance in Metals and Alloys*, ed: American Society for Testing and Materials, 1970, pp. 53-76.

[55] R. Chaouadi, "Flow and Fracture Behavior of 9%Cr-Ferritic/Martensitic Steels," SCK CEN2005.

[56] G. Chen, Z.-S. Zhang, Y.-H. Mei, X. Li, D.-J. Yu, L. Wang, *et al.*, "Applying viscoplastic constitutive models to predict ratcheting behavior of sintered nanosilver lap-shear joint", *Mechanics of Materials,* vol. 72, pp. 61-71, 2014.

[57] G. Kang and Q. Gao, "Uniaxial and non-proportionally multiaxial ratcheting of U71Mn rail steel: experiments and simulations", *Mechanics of Materials,* vol. 34, pp. 809-820, 2002.

[58] J. L. Chaboche, "On some modifications of kinematic hardening to improve the description of ratchetting effects", *International Journal of Plasticity,* vol. 7, pp. 661-678, 1991.

[59] J. Lemaître and J.-L. Chaboche, *Mechanics of solid materials*: Press Syndicate of the University of Cambridge, 1990.

[60] J.-L. Chaboche, "Stress calculations for lifetime prediction in turbine blades", *International Journal of Solids and Structures,* vol. 10, pp. 473-481, 1974.

[61] J.-L. Chaboche, "Continuous damage mechanics - A tool to describe phenomena before crack initiation", *Nuclear Engineering and Design,* vol. 64, pp. 233-247, 1981.

[62] C. O. Frederick and P. J. Armstrong, "A mathematical representation of the multiaxial Bauschinger effect", *Materials at High Temperatures,* vol. 24, pp. 1-26, 2007.

[63] K. P. Walker, "Research and development program for nonlinear structural modelling with advanced time-temperature dependent constitutive relationships," United Technologies Research Center, NASA1981.

[64] P. O. Kettunen and V.-T. Kuokkala, *Plastic Deformation and Strain Hardening*, 2002.

[65] S. Guo, G. Kang, and J. Zhang, "A cyclic visco-plastic constitutive model for time-dependent ratchetting of particle-reinforced metal matrix composites", *International Journal of Plasticity,* vol. 40, pp. 101-125, 2013.

[66] M. Abdel-Karim, "Modified kinematic hardening rules for simulations of ratchetting", *International Journal of Plasticity,* vol. 25, pp. 1560-1587, 2009.

[67] R. L. Klueh and A. T. Nelson, "Ferritic/martensitic steels for next-generation reactors", *Journal of Nuclear Materials,* vol. 371, pp. 37-52, 2007.

[68] K. Laha, S. Saroja, A. Moitra, R. Sandhya, M. D. Mathew, T. Jayakumar, *et al.*, "Development of India-specific RAFM steel through optimization of tungsten and tantalum contents for better combination of impact, tensile, low cycle fatigue and creep properties", *Journal of Nuclear Materials,* vol. 439, pp. 41-50, 2013.

[69] Q. Huang, C. Li, Y. Li, M. Chen, M. Zhang, L. Peng, *et al.*, "Progress in development of China Low Activation Martensitic steel for fusion application", *Journal of Nuclear Materials,* vol. 367-370, pp. 142-146, 2007.

[70] B. Fusion Engineering and DesignRaj and T. Jayakumar, "Development of Reduced Activation Ferritic–Martensitic Steels and fabrication technologies for Indian test blanket module", *Journal of Nuclear Materials,* vol. 417, pp. 72-76, 2011.

[71] R. L. Klueh, "Reduced-activation bainitic and martensitic steels for nuclear fusion applications", *Current Opinion in Solid State and Materials Science,* vol. 8, pp. 239-250, 2004.

[72] E. E. Bloom, "The challenge of developing structural materials for fusion power systems", *Journal of Nuclear Materials,* vol. 258-263, pp. 7-17, 1998.

[73] D. W. Kim and W.-S. Ryu, "Evaluation of Factors Affecting Cyclic Softening Behaviors of Mod. 9Cr-1Mo Steel", *Procedia Engineering,* vol. 10, pp. 376-382, 2011.

[74] J. F. Henry, "Growing experience with P91 T91 forcing essential code", *Combined Cycle Journal,* vol. First Quarter, 2005.

[75] "Welding "Grade 91" alloy steel", *V&M Tubes Corporation (Vallourec & Mannesmann),* vol. Inc. May, pp. 1-5, 2007.

[76] "Low Alloy Steel: P91 - Modified 9CrMo", *Metrode Products Data sheet.*

[77] W. B. Jones, C. R. Hills, and D. H. Polonis, "Microstructural Evolution of Modified 9Cr-1Mo Steel", *Metallurgical Transactions A,* vol. 22A, pp. 1049-1058, 1991.

[78] "Effect of Thickness on the Transformation Behavior of Grade 91 Steel," ASME Standards Technology, LLC2013.

[79] V. Karthik, K. Laha, P. Parameswaran, K. S. Chandravathi, K. V. Kasiviswanathan, T. Jayakumar, *et al.*, "Tensile properties of modified 9Cr-1Mo steel by shear punch testing and correlation with microstructures", *International Journal of Pressure Vessels and Piping,* vol. 88, pp. 375-383, 2011.

[80] V. Thomas Paul, S. Saroja, and M. Vijayalakshmi, "Microstructural stability of modified 9Cr–1Mo steel during long term exposures at elevated temperatures", *Journal of Nuclear Materials,* vol. 378, pp. 273-281, 2008.

[81] C. A. Hippsley and N. P. Haworth, "Hydrogen and temper embrittlement in 9Cr-1Mo steel", *Materials Science and Technology,* vol. 4, pp. 791-802, September 1988.

[82] V. Shankar, M. Valsan, K. B. S. Rao, R. Kannan, S. L. Mannan, and S. D. Pathak, "Low cycle fatigue behavior and microstructural evolution of modified 9Cr–1Mo ferritic steel", *Materials Science and Engineering: A,* vol. 437, pp. 413-422, 2006.

[83] V. Shankar, R. Sandhya, and M. D. Mathew, "Creep–fatigue-oxidation interaction in Grade 91 steel weld joints for high temperature applications", *Materials Science and Engineering: A,* vol. 528, pp. 8428-8437, 2011.

[84] B. Fournier, M. Sauzay, C. Caes, M. Noblecourt, M. Mottot, A. Bougault, *et al.*, "Creep–fatigue–oxidation interactions in a 9Cr–1Mo martensitic steel. Part I: Effect of tensile holding period on fatigue lifetime", *International Journal of Fatigue,* vol. 30, pp. 649-662, 2008.

[85] P. Zhao and F. Xuan, "Microstructural stability and its effect on ratchetting effect of advanced 9-12%Cr ferrite steel at high temperature", *Procedia Engineering,* vol. 27, pp. 1588-1595, 2012.

[86] A. Kumar, K. Laha, T. Jayakumar, K. B. S. Rao, and B. Raj, "Comprehensive microstructural characterization in modified 9Cr-1Mo ferritic steel by ultrasonic measurements", *Metallurgical and Materials Transactions A,* vol. 33A, pp. 1617-1626, 2002.

[87] A. Kumar, B. K. Choudhary, K. Laha, T. Jayakumar, K. B. S. Rao, and B. Raj, "Characterisation of microstructure in 9%Cr ferritic steel using ultrasonic measurements", vol. 56, pp. 483-497, 2003.

[88] C. G. Panait, W. Bendick, A. Fuchsmann, A. F. Gourgues-Lorenzon, and J. Besson, "Study of the microstructure of the Grade 91 steel after more than 100,000 h of creep exposure at 600 °C", *International Journal of Pressure Vessels and Piping,* vol. 87, pp. 326-335, 2010.

[89] G. Golański and S. Mroziński, "Low cycle fatigue and cyclic softening behaviour of martensitic cast steel", *Engineering Failure Analysis,* vol. 35, pp. 692-702, 2013.

[90] V. Shankar, K. Mariappan, R. Sandhya, and M. D. Mathew, "Evaluation of Low Cycle Fatigue Damage in Grade 91 Steel Weld Joints for High Temperature Applications", *Procedia Engineering,* vol. 55, pp. 128-135, 2013.

[91] M. Sauzay, H. Brillet, I. Monnet, M. Mottot, F. Barcelo, B. Fournier, *et al.*, "Cyclically induced softening due to low-angle boundary annihilation in a martensitic steel", *Materials Science And Engineering a-Structural Materials Properties Microstructure And Processing,* vol. 400, pp. 241-244, Jul 25 2005.

[92] S. Kim and J. R. Weertman, "Investigation of microstructural changes in a ferritic steel caused by high temperature fatigue", *Metallurgical Transactions A,* vol. 19A April, pp. 999-1007, 1988.

[93] R. Kannan, V. Sankar, R. Sandhya, and M. D. Mathew, "Comparative Evaluation of the Low Cycle Fatigue Behaviours of P91 and P92 Steels", *Procedia Engineering,* vol. 55, pp. 149-153, 2013.

[94] B. Fournier, M. Sauzay, F. Barcelo, E. Rauch, A. Renault, T. Cozzika, *et al.*, "Creep-Fatigue Interactions in a 9 Pct Cr-1 Pct Mo Martensitic Steel: Part II. Microstructural Evolutions", *Metallurgical and Materials Transactions A,* vol. 40, pp. 330-341, 2009.

[95] B. Fournier, M. Sauzay, C. Caës, M. Noblecourt, M. Mottot, L. Allais, *et al.*, "Creep-Fatigue Interactions in a 9 Pct Cr-1 Pct Mo Martensitic Steel: Part I. Mechanical Test Results", *Metallurgical and Materials Transactions A,* vol. 40, pp. 321-329, 2009.

[96] M. F. Giordana, P. F. Giroux, I. Alvarez-Armas, M. Sauzay, and A. Armas, "Micromechanical modeling of the cyclic softening of EUROFER 97 steel", *Procedia Engineering,* vol. 10, pp. 1268-1273, 2011.

[97] J. F. Stubbins and D. S. Gelles, "Fatigue performance and cyclic softening of F82H, a ferritic-martensitic steel", *Journal of Nuclear Materials,* vol. 233-237, pp. 331-335, 1996.

[98] A. F. Armas, I. Alvarez-Armas, C. Petersen, M. Avalos, and R. Schmitt, "Internal and effective stress analysis during cyclic softening of F82H mod. Martensitic stainless steel", *European Structural Integrity Society,* vol. 29, pp. 45-51, 2002.

[99] H. J. Chang and J. J. Kai, "Effects of temperature on the low cycle fatigue and microstructures of HT-9 ferritic steel", *Journal of Nuclear Materials,* vol. 191-194, pp. 836-840, 1992.

[100] A. F. Armas, M. Avalos, I. Alvarez-Armas, C. Petersen, and R. Schmitt, "Dynamic strain ageing evidences during low cycle fatigue deformation in ferrtic-martensitic stainless steels", *Journal of Nuclear Materials,* vol. 258-263, pp. 1204-1208, 1998.

[101] A. A. Saad, W. Sun, T. H. Hyde, and D. W. J. Tanner, "Cyclic softening behaviour of a P91 steel under low cycle fatigue at high temperature", *Procedia Engineering,* vol. 10, pp. 1103-1108, 2011.

[102] B. Fournier, M. Sauzay, and A. Pineau, "Micromechanical model of the high temperature cyclic behavior of 9–12%Cr martensitic steels", *International Journal of Plasticity,* vol. 27, pp. 1803-1816, 2011.

[103] J. Jung, "A note on the influence of hydrostatic pressure on dislocations", *Philosophical Magazine A,* vol. 43, pp. 1057-1061, 1981.

[104] J. Casey and T. D. Sullivan, "Pressure dependency, strength-differential effect, and plastic volume expansion in metals", *International Journal of Plasticity,* vol. 1, pp. 39-61, 1985.

[105] D. C. Drucker, "Plasticity Theory, Strength-Differential (SD)Phenomenon, and Volume Expansion in Metals and Plastics", *Metallurgical Transactions,* vol. 4, pp. 667-673, March 1973.

[106] K. Kimura, H. Kushima, and K. Sawada, "Long-term creep deformation property of modified 9Cr–1Mo steel", *Materials Science And Engineering a-Structural Materials Properties Microstructure And Processing,* vol. 510-511, pp. 58-63, 2009.

[107] J. Fred, "MINUIT Tutorial-Function Minimization," CERN2004.

Schriftenreihe des Instituts für Angewandte Materialien

ISSN 2192-9963

Band 1 Prachai Norajitra
Divertor Development for a Future Fusion Power Plant. 2011
ISBN 978-3-86644-738-7

Band 2 Jürgen Prokop
Entwicklung von Spritzgießsonderverfahren zur Herstellung von Mikrobauteilen durch galvanische Replikation. 2011
ISBN 978-3-86644-755-4

Band 3 Theo Fett
New contributions to R-curves and bridging stresses – Applications of weight functions. 2012
ISBN 978-3-86644-836-0

Band 4 Jérôme Acker
Einfluss des Alkali/Niob-Verhältnisses und der Kupferdotierung auf das Sinterverhalten, die Strukturbildung und die Mikrostruktur von bleifreier Piezokeramik $(K_{0,5}Na_{0,5})NbO_3$. 2012
ISBN 978-3-86644-867-4

Band 5 Holger Schwaab
Nichtlineare Modellierung von Ferroelektrika unter Berücksichtigung der elektrischen Leitfähigkeit. 2012
ISBN 978-3-86644-869-8

Band 6 Christian Dethloff
Modeling of Helium Bubble Nucleation and Growth in Neutron Irradiated RAFM Steels. 2012
ISBN 978-3-86644-901-5

Band 7 Jens Reiser
Duktilisierung von Wolfram. Synthese, Analyse und Charakterisierung von Wolframlaminaten aus Wolframfolie. 2012
ISBN 978-3-86644-902-2

Band 8 Andreas Sedlmayr
Experimental Investigations of Deformation Pathways in Nanowires. 2012
ISBN 978-3-86644-905-3

Die Bände sind unter www.ksp.kit.edu als PDF frei verfügbar oder als Druckausgabe bestellbar.

Band 9 Matthias Friedrich Funk
Microstructural stability of nanostructured fcc metals during cyclic deformation and fatigue. 2012
ISBN 978-3-86644-918-3

Band 10 Maximilian Schwenk
Entwicklung und Validierung eines numerischen Simulationsmodells zur Beschreibung der induktiven Ein- und Zweifrequenzrandschichthärtung am Beispiel von vergütetem 42CrMo4. 2012
ISBN 978-3-86644-929-9

Band 11 Matthias Merzkirch
Verformungs- und Schädigungsverhalten der verbundstranggepressten, federstahldrahtverstärkten Aluminiumlegierung EN AW-6082. 2012
ISBN 978-3-86644-933-6

Band 12 Thilo Hammers
Wärmebehandlung und Recken von verbundstranggepressten Luftfahrtprofilen. 2013
ISBN 978-3-86644-947-3

Band 13 Jochen Lohmiller
Investigation of deformation mechanisms in nanocrystalline metals and alloys by in situ synchrotron X-ray diffraction. 2013
ISBN 978-3-86644-962-6

Band 14 Simone Schreijäg
Microstructure and Mechanical Behavior of Deep Drawing DC04 Steel at Different Length Scales. 2013
ISBN 978-3-86644-967-1

Band 15 Zhiming Chen
Modelling the plastic deformation of iron. 2013
ISBN 978-3-86644-968-8

Band 16 Abdullah Fatih Çetinel
Oberflächendefektausheilung und Festigkeitssteigerung von niederdruckspritzgegossenen Mikrobiegebalken aus Zirkoniumdioxid. 2013
ISBN 978-3-86644-976-3

Band 17 Thomas Weber
Entwicklung und Optimierung von gradierten Wolfram/EUROFER97-Verbindungen für Divertorkomponenten. 2013
ISBN 978-3-86644-993-0

Band 18 Melanie Senn
Optimale Prozessführung mit merkmalsbasierter Zustandsverfolgung. 2013
ISBN 978-3-7315-0004-9

Band 19 Christian Mennerich
Phase-field modeling of multi-domain evolution in ferromagnetic shape memory alloys and of polycrystalline thin film growth. 2013
ISBN 978-3-7315-0009-4

Band 20 Spyridon Korres
On-Line Topographic Measurements of Lubricated Metallic Sliding Surfaces. 2013
ISBN 978-3-7315-0017-9

Band 21 Abhik Narayan Choudhury
Quantitative phase-field model for phase transformations in multi-component alloys. 2013
ISBN 978-3-7315-0020-9

Band 22 Oliver Ulrich
Isothermes und thermisch-mechanisches Ermüdungsverhalten von Verbundwerkstoffen mit Durchdringungsgefüge (Preform-MMCs). 2013
ISBN 978-3-7315-0024-7

Band 23 Sofie Burger
High Cycle Fatigue of Al and Cu Thin Films by a Novel High-Throughput Method. 2013
ISBN 978-3-7315-0025-4

Band 24 Michael Teutsch
Entwicklung von elektrochemisch abgeschiedenem LIGA-Ni-Al für Hochtemperatur-MEMS-Anwendungen. 2013
ISBN 978-3-7315-0026-1

Band 25 Wolfgang Rheinheimer
Zur Grenzflächenanisotropie von $SrTiO_3$. 2013
ISBN 978-3-7315-0027-8

Band 26 Ying Chen
Deformation Behavior of Thin Metallic Wires under Tensile and Torsional Loadings. 2013
ISBN 978-3-7315-0049-0

Band 36 Christoph Hage
Grundlegende Aspekte des 2K-Metallpulverspritzgießens. 2014
ISBN 978-3-7315-0217-3

Band 37 Bartłomiej Albi´nski
Instrumentierte Eindringprüfung bei Hochtemperatur für die Charakterisierung bestrahlter Materialien. 2014
ISBN 978-3-7315-0221-0

Band 38 Tim Feser
Untersuchungen zum Einlaufverhalten binärer alpha-Messinglegierungen unter Ölschmierung in Abhängigkeit des Zinkgehaltes. 2014
ISBN 978-3-7315-0224-1

Band 39 Jörg Ettrich
Fluid Flow and Heat Transfer in Cellular Solids. 2014
ISBN 978-3-7315-0241-8

Band 40 Melanie Syha
Microstructure evolution in strontium titanate Investigated by means of grain growth simulations and x-ray diffraction contrast tomography experiments. 2014
ISBN 978-3-7315-0242-5

Band 41 Thomas Haas
Mechanische Zuverlässigkeit von gedruckten und gasförmig abgeschiedenen Schichten auf flexiblem Substrat. 2014
ISBN 978-3-7315-0250-0

Band 42 Aron Kneer
Numerische Untersuchung des Wärmeübertragungsverhaltens in unterschiedlichen porösen Medien. 2014
ISBN 978-3-7315-0252-4

Band 43 Manuel Feuchter
Investigations on Joule heating applications by multiphysical continuum simulations in nanoscale systems. 2014
ISBN 978-3-7315-0261-6

Band 44 Alexander Vondrous
Grain growth behavior and efficient large scale simulations of recrystallization with the phase-field method. 2014
ISBN 978-3-7315-0280-7

Band 45 Tobias Kennerknecht
Fatigue of Micro Molded Materials – Aluminum Bronze and Yttria Stabilized Zirconia. 2014
ISBN 978-3-7315-0293-7

Band 46 Christopher Scherr
Elektrochemisches Verhalten von Lithium-Schwefel-Zellen mit unterschiedlicher Kathodenstruktur. 2015
ISBN 978-3-7315-0296-8

Band 47 Konstantin Frölich
Der Decal-Prozess zur Herstellung katalysatorbeschichteter Membranen für PEM-Brennstoffzellen. 2015
ISBN 978-3-7315-0334-7

Band 48 Benedikt Haspel
Werkstoffanalytische Betrachtung der Eigenschaften von mittels neuartiger RTM-Fertigungsprozesse hergestellten glasfaserverstärkten Polymerverbunden. 2015
ISBN 978-3-7315-0337-8

Band 49 Marco Berghoff
Skalenübergreifende Modellierung und Optimierung vom atomistischen kristallinen Phasenfeldmodell bis zur mesoskopischen Phasenfeldmethode. 2015
ISBN 978-3-7315-0416-0

Band 50 Michael Selzer
Mechanische und Strömungsmechanische Topologieoptimierung mit der Phasenfeldmethode. 2016
ISBN 978-3-7315-0431-3

Band 51 Michael Mahler
Entwicklung einer Auswertemethode für bruchmechanische Versuche an kleinen Proben auf der Basis eines Kohäsivzonenmodells. 2016
ISBN 978-3-7315-0441-2

Band 52 Christoph Bohnert
Numerische Untersuchung des Verformungs- und Bruchverhaltens von einkristallinem Wolfram auf mikroskopischer Ebene. 2016
ISBN 978-3-7315-0444-3

Band 53 Stefan Guth
Schädigung und Lebensdauer von Nickelbasislegierungen unter thermisch-mechanischer Ermüdungsbeanspruchung bei verschiedenen Phasenlagen. 2016
ISBN 978-3-7315-0445-0

Band 54 Markus Klinsmann
The Effects of Internal Stress and Lithium Transport on Fracture in Storage Materials in Lithium-Ion Batteries. 2016
ISBN 978-3-7315-0455-9

Band 55 Thomas Straub
Experimental Investigation of Crack Initiation in Face-Centered Cubic Materials in the High and Very High Cycle Fatigue Regime. 2016
ISBN 978-3-7315-0471-9

Band 56 Maren Lepple
Kupfer- und Eisenoxide als Konversions-Elektrodenmaterialien für Lithium-Ionen-Batterien: Thermodynamische und Elektrochemische Untersuchungen. 2016
ISBN 978-3-7315-0482-5

Band 57 Stefan Andreas Slaby
Charakterisierung und Bewertung der Zug- und Ermüdungseigenschaften von Mikrobauteilen aus 17-4PH Edelstahl. Ein Vergleich von mikropulverspritzgegossenem und konventionell hergestelltem Material. 2017
ISBN 978-3-7315-0484-9

Band 58 Kumar Ankit
Phase-field modeling of microstructural pattern formation in alloys and geological veins. 2016
ISBN 978-3-7315-0491-7

Band 59 Kuo Zhang
Characterization and Modeling of the Ratcheting Behavior of the Ferritic-Martensitic Steel P91. 2017
ISBN 978-3-7315-0503-7